Stars and Planets

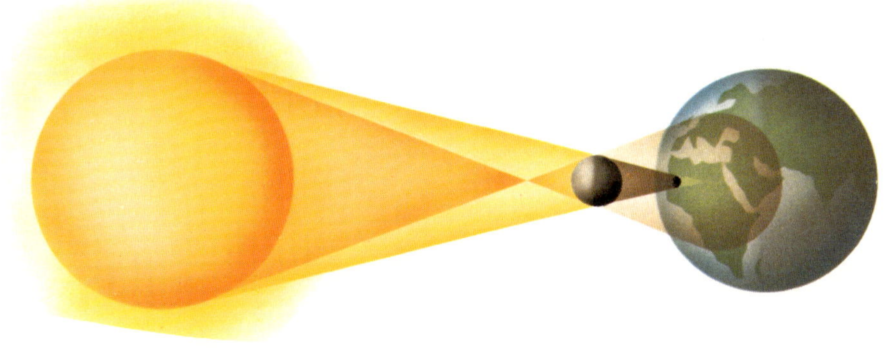

WARWICK PRESS

Contents

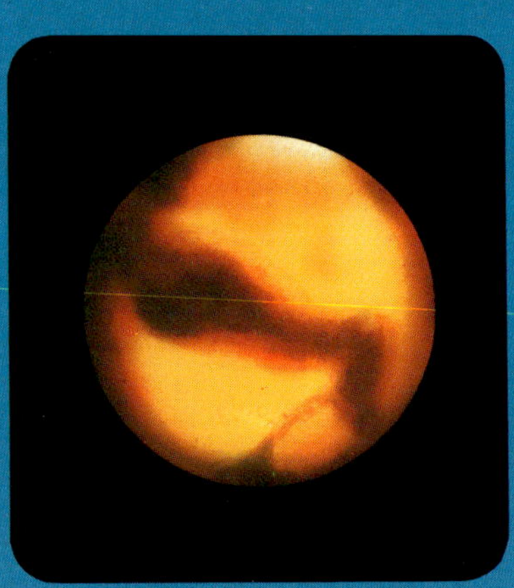

Left, top to bottom:
The Whirlpool galaxy.
Mars – the red planet.
A Saturn V rocket lifts Apollo 11
from the launching pad at the start
of its journey to the moon.

Author Keith Wicks
Editor John Paton

Published by Warwick Press, 730 Fifth Avenue, New York, New York 10019
First published in Great Britain by Sampson Low in 1977
Copyright © 1977 by Grisewood & Dempsey Ltd.
Printed in Great Britain by Purnell and Sons Ltd., Paulton (Avon) and London

Library of Congress Catalog Card No. 76–45483
ISBN 0–531–09077–9
ISBN 0–531–09052–3 lib. bdg.

Stars and Planets

Stars, planets, and other heavenly bodies fascinated ancient peoples. The movements of the stars across the skies were both useful and mysterious. The time of day could be judged from the position of the sun. But what made the heavenly bodies move as they did? Most people thought that the stars and planets must be controlled by supernatural powers. And many believed in astrology – the superstition that the positions of the stars somehow affect our lives. For centuries, astrology and astronomy were very much the same. Indeed, many of the things said by early astronomers were just as misleading as the forecasts of the astrologers. For example, the earth was once thought to be at the center of the universe. Only in the last few hundred years have we come to realize that the earth is a tiny, insignificant speck. Our small planet moves around the sun, which is just one of about 100,000 million stars in our Galaxy (star system). And at least 1,000 million other galaxies are known to exist!

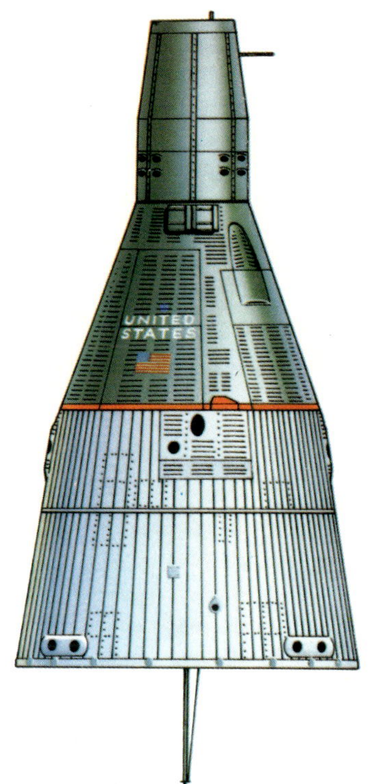

Above: A Gemini spacecraft.
Below: Mariner 10 took the first close-up pictures of Mercury in March 1974.

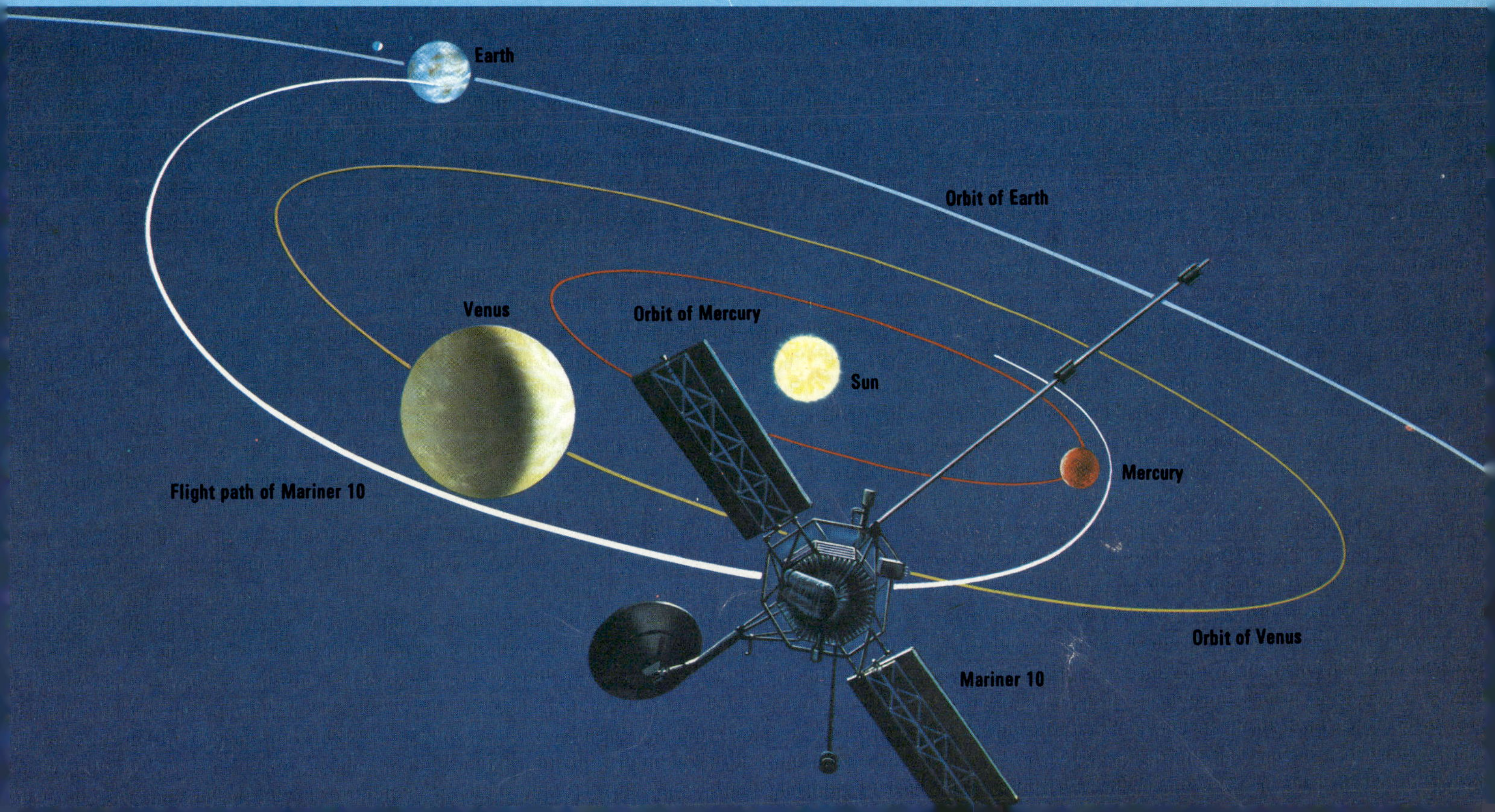

The Dawn of Astronomy

Early man explained the movements of the stars and planets with myths and legends. The Chinese thought an eclipse of the sun happened because a dragon was trying to eat it.

The symbols carved on this Babylonian stone are of various heavenly bodies. Venus, the moon, and the sun are at the top. The stone is about 3,000 years old.

Astronomy is the oldest science of all. It was natural for early peoples to study the motions of the heavenly bodies. The sun, moon and stars affected the way they organized their lives. In the light and warmth of day, man could hunt and carry out other tasks. So his activities were ruled by the times of sunrise and sunset. The moon was important to early man too, but not just because it gave light at night. The regular changes in the appearance of the moon – its *phases* – were a basis for the earliest calendars. It takes $29\frac{1}{2}$ days for the moon to go from full moon to full moon. This period of time is known as a lunar (moon) month.

The advantage of having a reliable calendar became clear around 8000 BC, when man started to farm the land. For, by dividing up the year, the sowing and harvesting of crops could be planned.

The ancient Babylonians thought that 12 lunar months were equal to a year. So they made a calendar with 12 months, alternately 29 and 30 days long. Unfortunately, this meant that their year was only 354 days long – about 11 days short of its true length. The error soon became obvious because, over the years, the months got more and more out of step with the seasons. To overcome this problem, an extra month was added every third year. The new system was fairly accurate. But the year on the calendar was still a little too short and further corrections had to be made from time to time. The Babylonians knew that certain patterns of stars became visible at certain times of the year. So, from about 2100 BC, they used these observations to check the calendar. Whenever a star appeared in the wrong month according to the calendar, a correction was made. In other countries, different calendar systems were made, and these, too, were based on astronomical observations of the moon and stars.

Myths, Legends, and Religion

With accurate calendars, astronomers were able to forecast future events such as eclipses and the appearance of comets. But they still could not explain why such events took place. And they had no idea of the true nature of the heavenly bodies. So myths and legends

An Egyptian coffin, about 2,000 years old, bears this painting of Nut, the sky goddess. The animals around her are symbols for constellations.

developed to explain the unknown. The Chinese, for example, thought that an eclipse of the sun took place because a dragon was trying to eat it. So they would all shout and beat gongs to scare away the dragon. These tactics seemed to work, for the sun soon appeared normal again.

The progress of astronomy was held up for a long time by the fact that many people worshiped the sun and moon as gods. Fortunately, a few people went on asking questions about what they saw, and astronomy slowly progressed.

The Earth and its Place

All the early astronomers thought that the earth was flat. The first suggestion that the earth was curved came from the Greek philosopher (thinker) Aristotle, in the 300s BC. A curved earth, he ex-

Above: Astronomers at Istanbul Observatory in the Middle Ages. Left: Ptolemy's idea of the universe. He thought that the sun, moon, and planets circled the earth. Below: How Ptolemy explained the movements of the planets. He said that they moved in small circles (epicycles), which moved in larger circles (deferents).

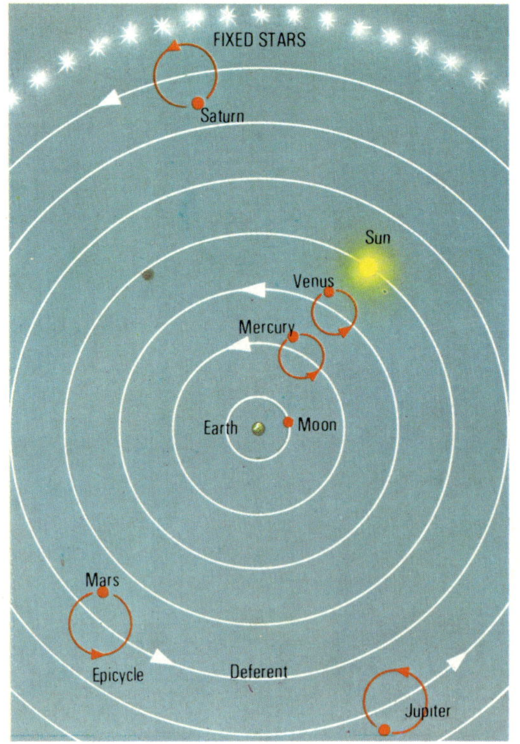

plained, would account for the fact that certain stars can be seen from one country, but not from another. And it would also account for the curved shadow of the earth that could be seen on the moon during an eclipse. It was soon accepted that the earth was, in fact, a ball. And, since our planet was the most important of all the heavenly bodies, the Greeks assumed that it was at the center of the universe. Around it, they said, circled the sun, moon, and the other planets, against a background of fixed stars. But this did not account for the strange movements of the planets. However, in the 100s AD, the Greek philosopher Ptolemy put forward the following explanation. The planets, he said, move in small circles (epicycles), and these move in large circles (deferents) around the earth. For 14 centuries, Ptolemy's incorrect system was accepted as the truth.

Above: Nicolaus Copernicus, who started a revolution in astronomy in the 1500s.

Right: The universe as Copernicus saw it. The planets move around the sun, and the moon moves around the earth.

Men Who Changed Astronomy

In the 1500s, Nicolaus Copernicus published a book in which he said that the planets revolve around the sun. This led other astronomers to make important new discoveries about the planets and their movements.

After the death of the Greek philosopher Ptolemy in 180 AD, little progress was made in astronomy. Then in the 700s, the Arabs made Baghdad a center for studying the heavens. Although astrology was still combined with astronomy, many careful observations were made. But the development of astronomy was slow because people still believed that the earth was the center of the universe. Not until the 1500s was it accepted that the planets orbit the sun. The man who started this revolution in astronomy was called Nicolaus Copernicus, a Polish doctor.

Johannes Kepler showed that the paths of the planets around the sun are ellipses, not circles.

The Copernican Revolution

Copernicus was not really the first to think that the planets moved around the sun. The Greek Aristarchus had the same idea in about 290 BC. But he failed to convince others, and the idea of an earth-centered universe therefore went on. When Copernicus realized, from his observations, that the planets move around the sun, he did not publish his findings at once. For the Church strongly supported Ptolemy's idea of the universe. And anyone opposing it was liable to be tortured or even executed. So Copernicus took the wise step of delaying the publication of his theory until 1543, when he knew he had not long to live. In spite of strong opposition from the Church the Copernican system slowly gained more supporters. The revolution had begun.

The Dane Tycho Brahe was an expert astronomer of the 1500s. But, although he recorded lots of very accurate information, he did not make full use of it. Brahe always believed in an earth-centered universe. However, his pupil, a German mathematician called Johannes Kepler, put the records to good use. From them, he was able to work out exactly how the planets move. In 1609, Kepler discovered that the planets move around the sun in ellipses, not circles. He also found that the planets move fastest when they are closest to the sun.

While Kepler was working out the laws which govern the motions of the planets, a most important event took place in Holland. In 1608, a spectacle maker called Hans Lippershey invented the telescope. The Italian Galileo Galilei promptly built one. With it

Galileo Galilei, the Italian who introduced the telescope to astronomy in the 1600s.

he discovered sunspots, the phases of Venus, the moon's mountains and valleys, and four of Jupiter's moons. But the Church still insisted that the earth-centered theory was correct, and made Galileo promise to stop supporting the Copernican system. Later, the Church forced Galileo to give up astronomy altogether.

Motion and Gravity

Isaac Newton, a brilliant English scientist, made many discoveries of great importance to astronomy. In the 1660s, he built the first reflecting telescope. Newton also worked out laws concerning bodies in motion, including falling bodies. This work led him to realize that a force of attraction must exist between any two objects. We call this force *gravitation*. It makes things fall to the ground, and also keeps the planets in their orbits around the sun. Newton's laws enabled astronomers to calculate the movement of heavenly bodies with great accuracy. For example, the next appearance of a comet could be forecast with confidence.

Thoughts on the Universe

One of the greatest astronomers of all time was William Herschel. From his observations of the stars, Herschel worked out the shape of our Galaxy. And, in the 1780s, he was the first to suggest that some objects visible in the night sky might be separate galaxies at immense distances from our own star system. If this proved to be true, it would mean that the universe was far more vast than had been supposed. Herschel's theory was proved in 1923 by the American scientist Edwin Hubble. Then man's ideas about the universe changed dramatically. And he was soon to discover that the universe itself was changing dramatically, too.

Above: Isaac Newton discovered that sunlight could be split into a range of colors called the spectrum.

William Herschel predicted that galaxies would be discovered outside our own star system.

The Origin of the Universe

Why the universe began is a complete mystery. But more and more scientists are agreeing about the way in which it started.

The universe is expanding! This dramatic announcement, made by the American astronomer Edwin Hubble in 1929, provided clues to the origin of the universe. From his observations of distant galaxies, (star systems), Hubble had found out that they were moving away from us in an orderly manner.

In many ways, the expanding universe resembles a spotted balloon being blown up. The spots, which represent groups of galaxies, get farther and farther apart as the balloon expands. As a result of Hubble's discovery, two main theories were developed to explain how the universe was formed.

Big Bang or Steady State?
The big bang theory was thought of in the 1930s by Georges Lamaître, a Belgian priest and astronomer. According to this theory, the universe started with an enormous explosion that sent material flying out in all directions. This material gradually formed into the galaxies.

In 1948, Hermann Bondi, Thomas Gold, and Fred Hoyle produced a new theory. They thought that the universe was in a "steady state". Although the distant galaxies were moving away, new matter was continuously being created. This kept the density of the universe constant.

Most scientists now support the big bang theory. They think that the bang — the creation of the universe — took place 12 to 25 thousand million years ago.

The universe probably began thousands of millions of years ago with an enormous explosion. This is usually referred to as the "big bang". Vast clouds of gases shot out, cooled gradually and formed the galaxies.

The Galaxies

Immense star systems called galaxies are the building blocks of the universe. Thousands of millions of galaxies are known to exist.

Our sun is just one of about 100,000 million stars in a vast, turning spiral that we call our Galaxy. On a clear night, we can see several thousand of these stars with the naked eye. In some parts of the sky, there are so many stars that they appear to form a white band. We call this band the Milky Way.

The distance across our Galaxy is about one million million million kilometers (600 thousand million million miles). Astronomers often express large distances like this in light years. A light

Above: Part of our Galaxy, the Milky Way. So many stars lie in this direction that they appear to form a cloudy, white band across the sky. A passing artificial satellite caused the thin, bright streak across the picture.

Right: Two views of our Galaxy, showing how the stars are arranged in a flat spiral. The distance across the Galaxy is about 100,000 light years. The red arrows show the position of our sun.

year is the distance that light travels through space in a year. Light travels at 300,000 kilometers per second (186,000 miles per second), and would take 100,000 years to cross our Galaxy. So the distance across the Galaxy is said to be 100,000 light years.

Besides stars, the spiral arms of the Galaxy contain clouds of hydrogen gas and *nebulae*. Nebulae are vast clouds of dust and gases. Astronomers believe that new stars are formed from the nebulae.

Right: A spiral galaxy. The center is quite distinct from the spiral arms.

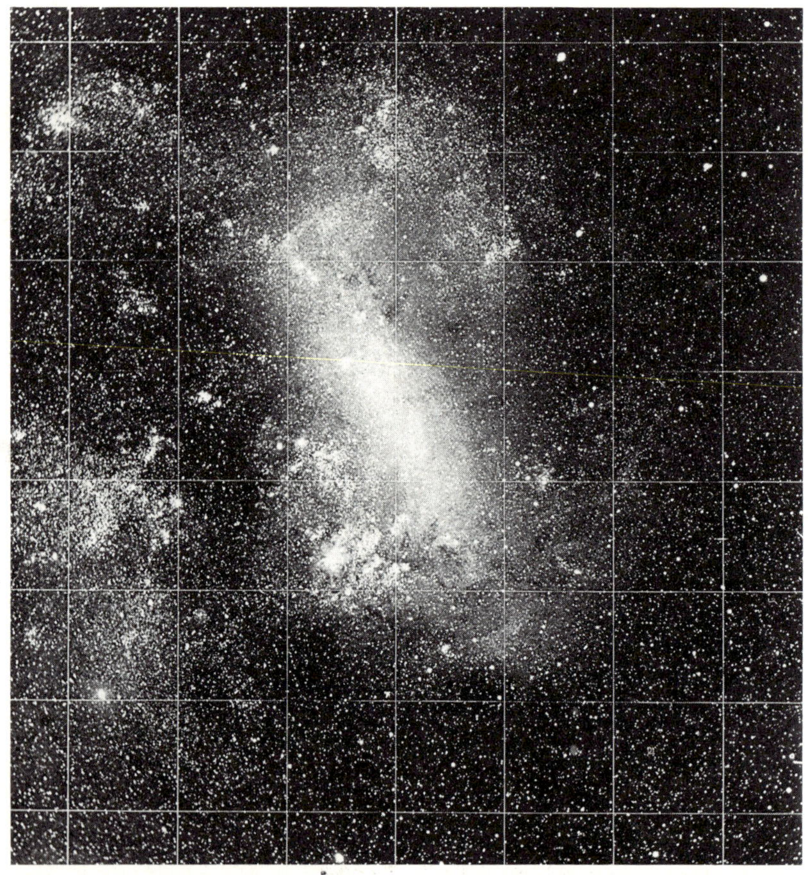

Below: The Large Magellanic Cloud is the closest galaxy to our own Milky Way. It is about 170,000 light years away.

Islands in Space

In the 1700s, William Herschel suggested that some of the so-called nebulae, although they appeared cloud-like, were in fact "island universes". These were complete star systems at enormous distances from our Galaxy. More and more evidence found by astronomers in the 1800s supported this theory. And, in 1917, the new giant telescope at Mount Wilson, California, showed that the Andromeda "nebula" was made up of stars. Herschel's theory was finally confirmed in 1923. Then Edwin Hubble proved that these stars were hundreds of thousands of light years from the earth. This meant that the Andromeda nebula was really a galaxy, quite separate from our own star system.

Thousands of millions of galaxies are now known to exist. Some have no definite shape; others are spiral or elliptical. The galaxies are arranged in groups. Our own Galaxy and 16 others within a distance of about 3 million light years are known as the Local Group.

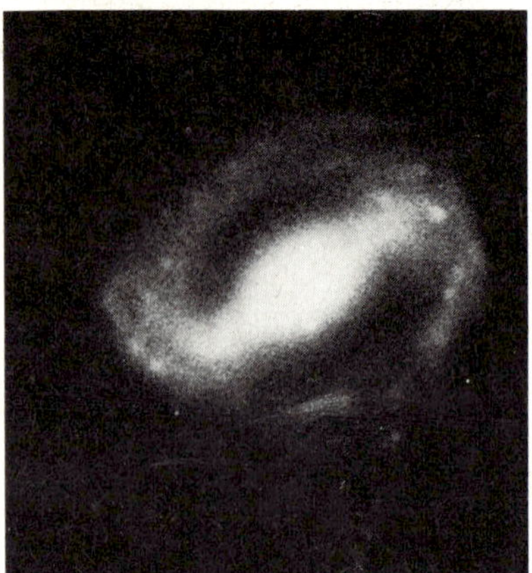

Left and far left: Two kinds of spiral galaxies. They differ from ordinary spiral galaxies in that the arms of the spiral trail from the ends of a bar-like formation passing through the center of the galaxy.

The Stars

Stars vary immensely in size and behavior. Some are smaller than the earth. Others are thousands of times larger. And, while most stars shine steadily, others flash on and off.

Because stars give off light and heat, astronomers once thought that they were burning. But this was shown to be wrong by William Thomson, the Scottish physicist who later became Lord Kelvin. He showed that, if the sun produced its energy by burning, it could not last for more than a few thousand years. As the sun was known to be much older than this, it could not be burning. And this presumably applied to the more distant stars too.

We now know that the stars produce their energy by turning hydrogen into helium. This process, which gives vast amounts of energy, is known as a *nuclear reaction*. The same reaction takes place in hydrogen bombs. But, in stars, the reaction takes place at

Above: Among the stars are huge clouds of glowing gas and dust called nebulae. Shown here is the Trifid nebula, which lies in the constellation Sagittarius. New stars are formed from clouds of this kind. Below: Tens of thousands of stars make up this cluster known as Hercules star cluster M13.

a steady rate. So the energy they give out remains almost constant
for millions of years.

Size and Brightness

Compared with the earth, the sun is enormous. But it is one of the
smallest stars and is called a *dwarf*. The largest stars, called *supergiants*, are up to 3,000 times the size of the sun.

The brightness (*magnitude*) of stars varies too. Around 150 BC,
the Greek astronomer Hipparchus worked out a system for
numbering stars according to their brightness. The brightest stars
were said to be of magnitude 1, and the faintest of magnitude 6. On
average, magnitude 1 stars were about 100 times brighter than stars
of magnitude 6. Today, we use a similar system, but it has been
extended to cover very bright and very faint stars. As a small magnitude number means a bright star, the brightest of all have been
given magnitude numbers of less than 1. Sirius, for example, has a
magnitude of −1·45. And the faintest stars, seen only with the
most powerful telescopes, have magnitudes of more than 20.

The Life History of a Star

Stars form in enormous clouds of dust and gas called nebulae. A
star starts to form when the force of gravity causes the gas and dust
in one part of a nebula to collect together. The strong force
squeezes the dust and gas into a huge ball. The force on each part
of the ball is directed toward its center. As a result, the pressure at
the center is enormous. Now, when the pressure of a gas increases,
its temperature increases too. (This is the reason why, when a
bicycle tire is pumped up, the pump and tire become warm.) So
the center of the ball becomes hot. As gravity squeezes the dust and

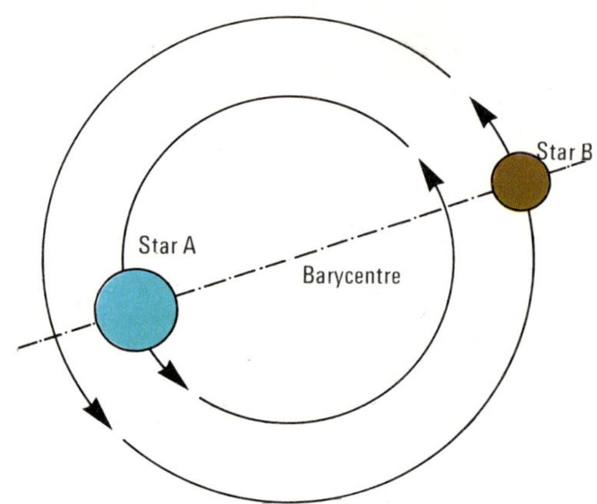

Above: In a typical binary star system, one of the
stars is bigger than the other. They both orbit
around a central point.

Below: Some binary star systems appear to
"wink". These are called eclipsing binaries. One
star is dimmer than the other. It regularly passes
directly between the brighter star and the earth.
This eclipses the brighter component, and reduces
the overall brightness of the binary pair for a short
time.

Here, the sun, a dwarf star, is compared in size
with giant and supergiant stars. The largest
supergiants are about 3,000 times the size of the
sun.

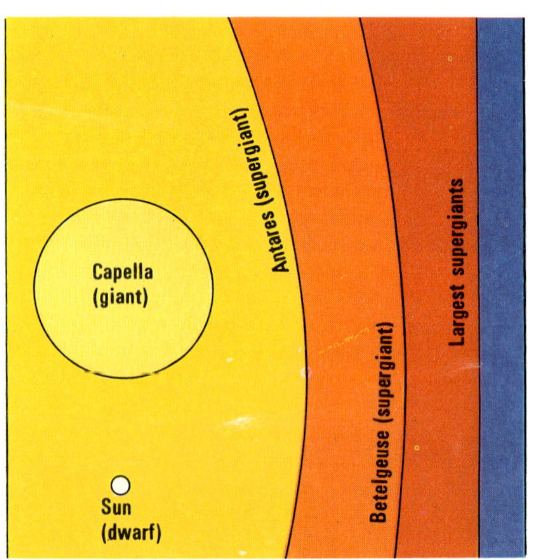

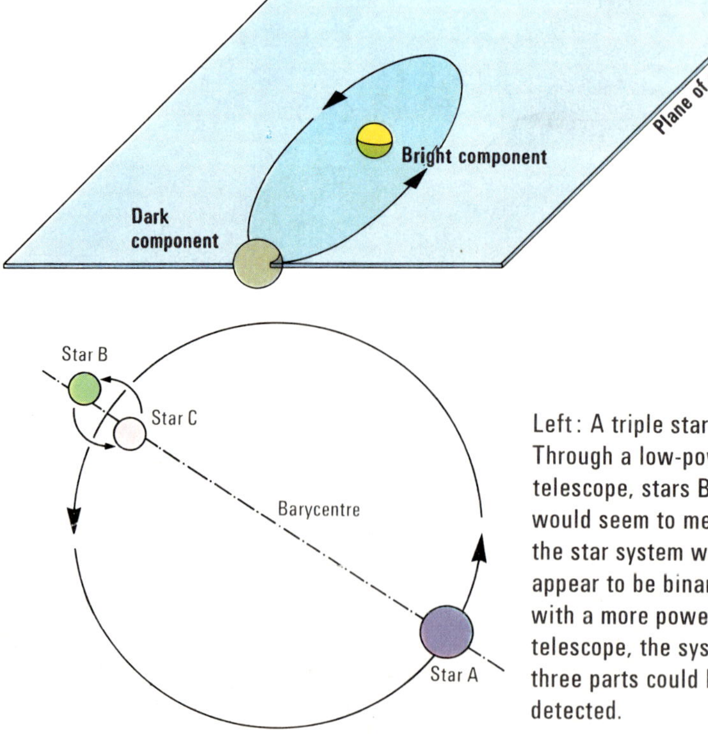

Left: A triple star system.
Through a low-powered
telescope, stars B and C
would seem to merge, and
the star system would
appear to be binary. But,
with a more powerful
telescope, the system's
three parts could be
detected.

gas into a smaller and smaller ball, so the pressure and temperature at the center rise. After a while, the ball glows red hot, and a star is born.

The temperature of the new star rises steadily as gravity squeezes it tighter and tighter. When the temperature passes about 10 million degrees, there is a big change in the star. Everything is right for nuclear reactions to take place. Hydrogen is turned into helium, and large amounts of energy are radiated. The star has now reached a settled stage, and stays like this for much of its life.

The largest of the new stars take about 100,000 years to settle down. Then, the nuclear reactions that take place make them glow with a brilliant white or blue light. This goes on for a few million years – quite a short time for stars.

Smaller new stars take longer to settle down. They become dim red or yellow dwarf stars. In these, the nuclear reactions take place much more slowly, so they have a much longer life. It may take as

long as 20 million million years for one of these dwarfs to use up all its hydrogen fuel.

When a star's hydrogen has been used up, new reactions start to take place. The temperature at the center of the star rises. And the star swells until it grows into a giant or supergiant. Other changes then take place. A large star may explode and become 100 million times brighter than the sun. A star exploding in this way is called a *supernova*. But most stars simply shrink from the giant stage to form small, dense stars called *white dwarfs*. These dwarfs, often no bigger than the earth, slowly cool down and fade out. Stars may linger in the white dwarf stage for millions of millions of years before they finally become extinct.

Star Groups

In ancient times, men noticed that the stars form patterns in the sky. They named these patterns, or *constellations*, after various gods, heroes, animals, and objects. The stars in a constellation look close together because they lie in roughly the same direction. In fact, some stars are thousands of light years farther away than their apparent neighbors. Although most stars do have close companions, they may not be visible to the naked eye. A group of stars that can be seen is the *cluster* (star group) called Pleiades, or the Seven Sisters. This cluster is in the constellation Taurus, the Bull. Only a few stars belonging to Pleiades can be seen with the naked eye. But this cluster actually contains more than 100 stars.

Many stars are in close pairs. Each star of the pair revolves around the same point. An arrangement of this kind is called a *binary star* system. Some binaries can be seen with the naked eye. But most appear as one star, and many cannot be separated with even the best telescope. But the kind of light that they give out tells astronomers that they are, in fact, binaries.

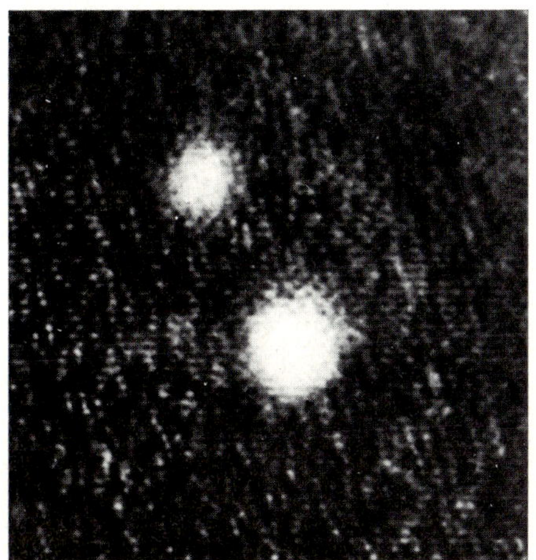
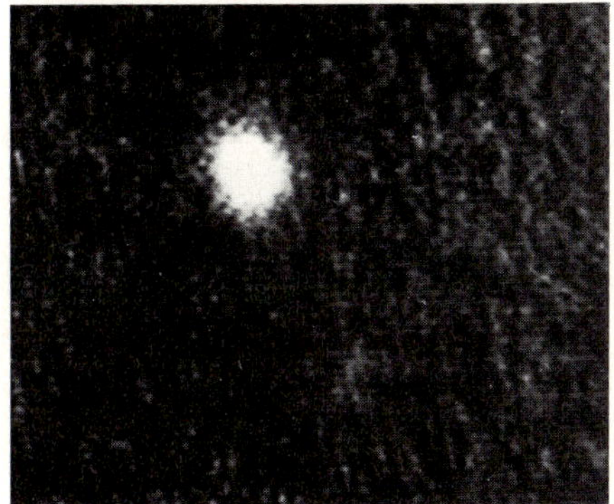

Below: These two photographs show the same part of the night sky. The larger star in the top picture is the Crab pulsar, or pulsating star. Pulsars radiate energy in short bursts. In the lower picture, the pulsar was "off", and did not register on the film.

The Solar System

The planets and the sun are linked by the attraction of gravity. They make up what we call the solar system.

No one knows how many planets there are moving around (orbiting) the sun. Mercury, Venus, Mars, Jupiter, and Saturn were known to the ancient astronomers. To them, these planets seemed to be stars. But they were unusual because they moved across the sky. For this reason, they were known as the

Name	Mean distance from Sun (in millions km mi)		Equatorial diameter (km) (mi)		Density (water=1)	Circles Sun in	Turns on axis in
Sun	—		1,392,000	865,000	1·41	—	25·38d
Moon	—		3476	2160	3·34	—	27·32d
Mercury	58	36	4850	3015	5·4	88d	59d
Venus	108	67	12,140	7545	5·2	224·7d	244d
Earth	150	93	12,756	7926	5·52	365·25d	23:56h
Mars	228	142	6790	4220	3·95	687d	24:37h
Jupiter	778	484	142,600	88,600	1·34	11·9y	9:50h
Saturn	1427	887	120,200	74,700	0·70	29·5y	10:14h
Uranus	2870	1783	49,000	30,500	1·58	84·0y	10:49h
Neptune	4497	2794	50,200	31,200	2·30	164·8y	15:48h
Pluto	5900	3670	6400	3980	?	247·7y	153h

wanderers. Our word "planet" comes from the Greek for "wanderer".

By the 1700s, astronomers knew that the planets were quite different from the stars. They realized that planets could be seen only by reflected sunlight. Stars gave out their own light. And they had worked out the orbits of the known planets quite accurately. But no new planets had been discovered since ancient times. However, in the 1600s, a German astronomer called Johannes Kepler had forecast that a planet would be discovered somewhere between the orbits of Mars and Jupiter. For the distance between these two orbits was much greater than the distances between the orbits of Mercury, Venus, Earth, and Mars. In 1801, Kepler's idea was proved correct. A tiny planet was discovered by Giuseppe Piazzi. The new planet was only 687 kilometers (427 miles) in diameter. The existence of such a small planet was a great surprise, and astronomers began to wonder if others might be found. How right they were! Thousands of these *minor planets*, or *asteroids*, have since been discovered.

The last of the nine known *main planets* to be discovered were Uranus (1781), Neptune (1846), and Pluto (1930). The search is now on for planet ten.

In the upper part of the diagram, the orbits, or paths, of the major planets are shown drawn to scale. The broad blue band between the orbits of Mars and Jupiter shows the orbits of the minor planets, or asteroids. And the orange line shows the orbit of a comet – a rare visitor to our part of the solar system. The diagram also shows the sun and planets drawn to scale.

The Sun

The sun is just one of 100,000 million stars that make up our Galaxy. Like other stars, it gives out energy produced by powerful nuclear reactions.

Our sun is a yellow dwarf, a small, quite dim star. It appears so large and bright because it is much closer than any other star – only 93 million miles (150 million kilometers) away. Like other stars, the sun gives out energy produced by nuclear reactions in its hot core. There, the temperature is around 15,000,000°C, compared with about 6,000°C at the sun's surface. In these reactions, about 600 million tons of hydrogen are turned into helium every second. This has been happening for about 5,000 million years. And enough hydrogen is left for the sun to continue in this way for another 5,000 million years.

The sun sends out its energy as *electromagnetic waves*. These include heat, radio

Above: A spectacular solar eruption photographed from Skylab space laboratory.

Right: An eclipse of the sun is seen when the moon's shadow falls on the earth. A total eclipse is seen from regions on which the deep shadow (umbra) falls.

Below: A series of pictures showing a solar prominence – a violent eruption at the sun's surface.

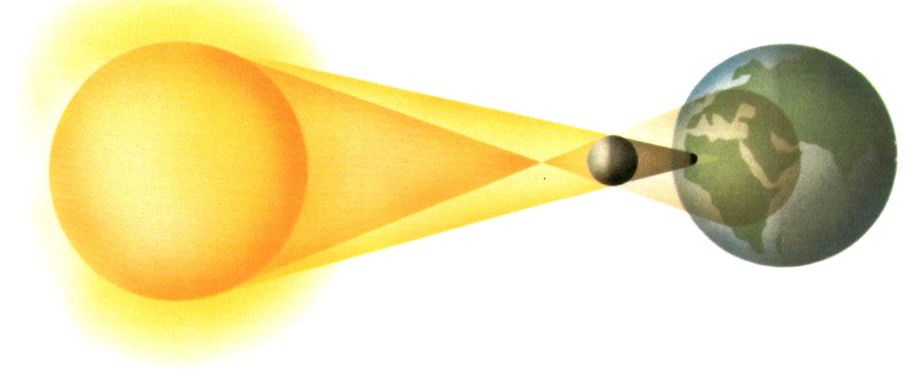

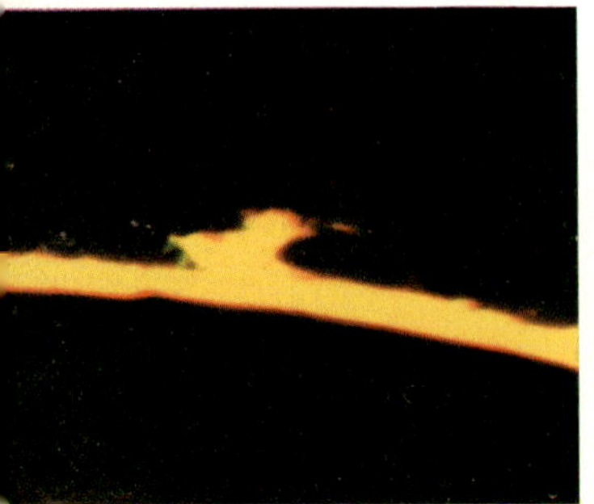

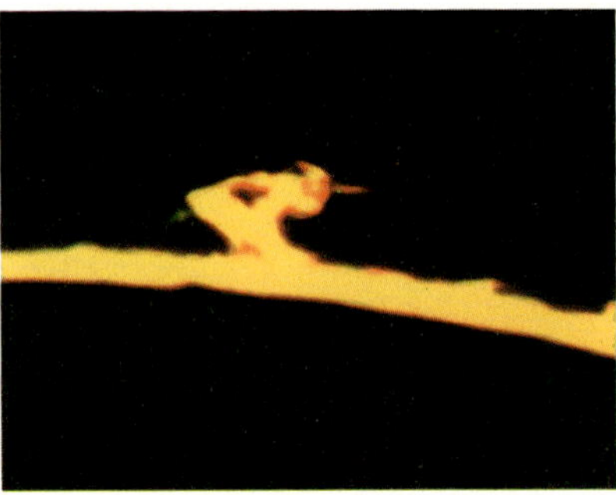

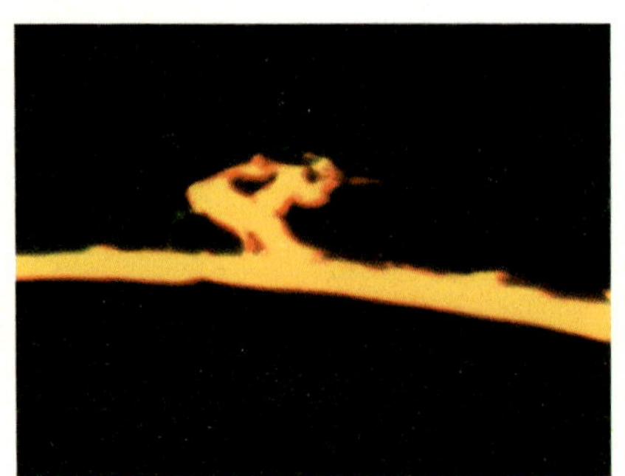

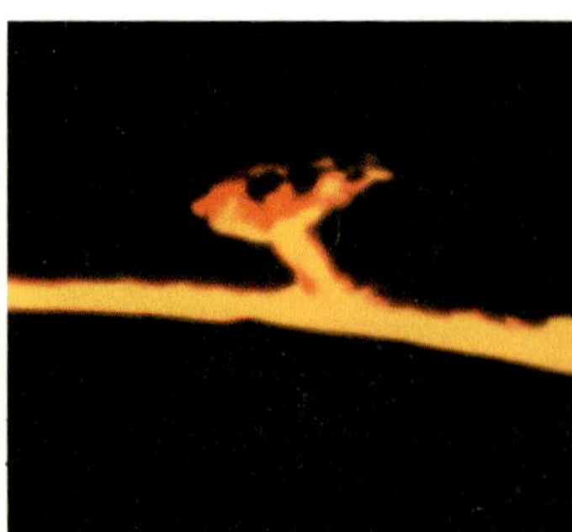

waves, light, ultraviolet rays, X-rays, and gamma rays. Only a tiny fraction of this energy reaches the earth. It gives us light, heat and almost all the energy we use.

The Solar Surface

Dark marks called *sunspots* sometimes appear on the sun's surface. Small spots

Photographs of the sun reveal dark patches called sunspots. These areas are about 2,000°C cooler than their surroundings. Therefore, although extremely bright, they are much dimmer than the rest of the sun. This is why they appear to be black in photographs.

Left: A total eclipse of the sun, showing its corona, or outer atmosphere. The sun has to be photographed through a dark glass to reduce the brightness of the image on the film. This makes the surrounding sky appear dark in comparison.

may last only a few hours. But larger ones sometimes last for months. Exactly how these spots form is uncertain. But they seem to have something to do with *flares*. These flares are extremely bright eruptions which shoot out streams of electrically charged particles from the sun's surface. The stream of particles called the *solar wind*, can interfere with radio communications on earth. And it sometimes causes displays of light, called auroras, at the earth's poles.

The biggest disturbances on the sun's surface are called *prominences*. Blazing streams of hydrogen shoot out from the sun. They sometimes reach heights of several hundred thousand miles. Prominences can be seen clearly during a total eclipse of the sun.

Solar Eclipses

An eclipse of the sun happens when the moon passes between the sun and earth. In regions where the moon blocks out all sunlight, the eclipse is said to be *total*. But, from many places, the moon does not appear to pass across the center of the sun. So only a *partial* eclipse is seen.

Left: Just before and after the sun is totally eclipsed, a small part of it is visible over the edge of the moon. This gives rise to the beautiful 'diamond-ring' effect shown here.

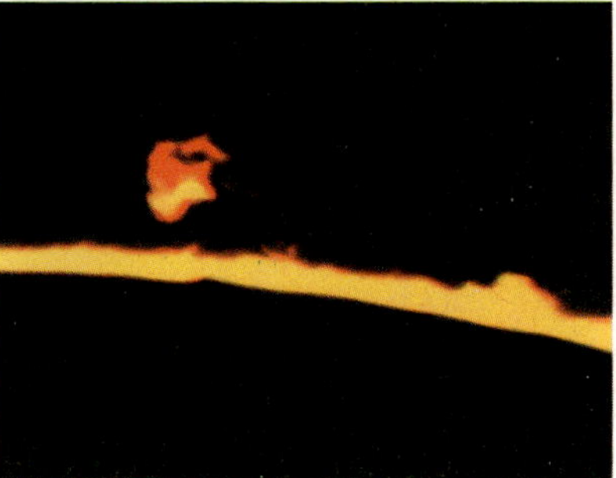
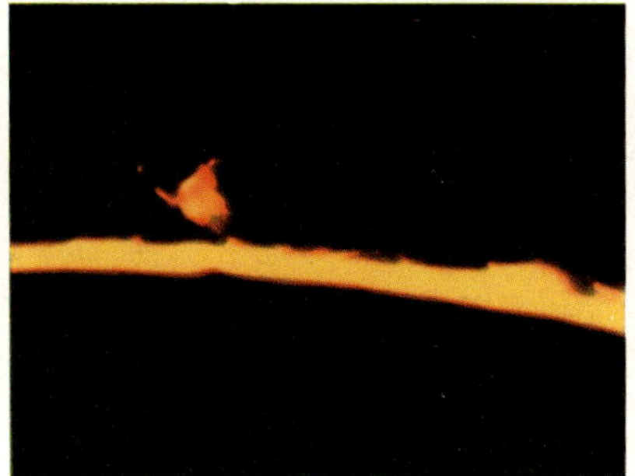
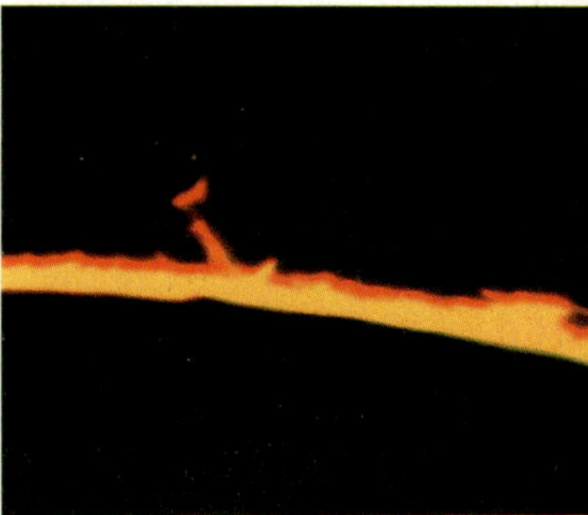

The Earth

No one knows for certain how the earth was formed. But scientists have discovered much about the way it developed from a ball of molten rock to its present state.

Above: The earth, photographed from Apollo 17 in December, 1972.

Right: Below the earth's thin crust is a much thicker layer of solid rock, called the mantle. Below this is the core.

Left: The Grand Canyon in Arizona. Over millions of years, the Colorado River has gradually worn away the land.

Bottom left: This Icelandic island was formed by a volcanic eruption.

From space, the earth is a beautiful, bright globe. Seventy percent of its surface is covered with water, which reflects sunlight well. Between the brilliant white polar ice caps are deep blue oceans and brownish continents, partly covered by swirling patches of white clouds.

The Changing Earth

About 4,600 million years ago, our planet was formed. It was then a molten mass of red-hot rock. Over millions of years, the rock gradually cooled, and a solid crust started to form. But the crust often cracked as it cooled down. Molten rock from the interior then welled up through the cracks and solidified on the surface.

As the earth's crust cooled down, steam and other gases escaped from the interior. Most of the steam then condensed to form the earth's surface waters. And other gases remained above the surface

as the atmosphere. The earth's gravity prevented these gases from escaping into space.

About 1,000 million years ago, the surface of our planet was one island surrounded by a vast ocean. But stresses in the earth's crust slowly pulled the island apart to form separate continents. The east coast of South America and the west coast of Africa are so similar in shape because they were once joined together.

The shaping of the earth's surface happened quite late in its history. About 400 million years ago, movements in the crust caused parts of it to fold up and form the first mountain chains. But some mountains were formed much more recently. The Alps, Himalayas, and Rocky Mountains, for example, were formed only 25 million years ago.

And the earth is still slowly changing. The gradual separation of huge chunks of the crust goes on to this day. The sudden slipping of one section against another causes earthquakes. And, in places where the crust is weak, magma (molten rock) from below sometimes bursts through to the surface. The result is a volcano.

The Tides

The moon's gravity attracts the seas directly beneath it, causing them to bulge. And, as the moon pulls on the earth more than on the waters on its far side, a bulge occurs there too.

The sun affects the earth's waters too, but not to the same extent as the closer moon. The tides vary with the moon's phases. At the first and last quarters of the moon, the sun's gravity reduces the effect of the moon's gravity on the waters. So the lowest tides occur at these times (top diagram). And the highest tides occur at new and full moon (lower diagram).

The Seasons: The earth's axis is inclined to the plane of its orbit around the sun. So each hemisphere is alternately tilted towards and away from the sun. This gives rise to the seasons.

The Atmosphere

Nitrogen makes up about 78 percent of our atmosphere. Most of the remainder is oxygen (21 percent). But there are traces of water vapor, argon, carbon dioxide, neon, and other gases. Animal life is dependent on the atmosphere's oxygen. And the atmosphere absorbs harmful ultraviolet radiation from the sun. The atmosphere also prevents the earth from cooling down too much at night.

Life on Earth

No other planet in the solar system can support the many forms of life found on the earth. For only our planet has all the conditions needed for living things to develop to such an extent. How life began is uncertain. But it is unlikely that the chemical substances in living organisms could ever have formed if water had not been present. The first primitive organisms may have developed over 3,000 million years ago. Over millions of years, many more complex organisms developed. Scientists believe that this slow process, called *evolution*, led to the great variety of plants and animals on the earth today. Modern man appeared on the scene quite recently – less than 400,000 years ago.

The Moon

The moon is the earth's only natural satellite. Its gravitational attraction makes the oceans bulge. This causes the tides.

The moon is a quiet, lifeless world. At an average distance of 384,400 kilometers (238,900 miles) from the earth, it is our nearest neighbor in space. The low gravity at its surface – about one-sixth as strong as on the earth – is too small to hold an atmosphere around it. With no atmosphere to scatter sunlight, shadows on the moon are dense. At midday, the moon's surface temperature reaches 100°C. But, at night, it drops to 150°C below zero.

Left: The moon, photographed by US astronauts from their Apollo 8 spacecraft in December, 1968.

Below: The shape of the moon appears to alter as it makes its monthly journey around the earth. We call the various shapes we see the phases of the moon.

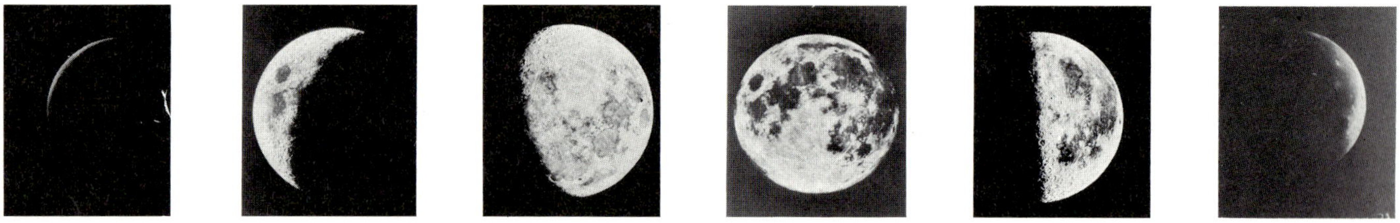

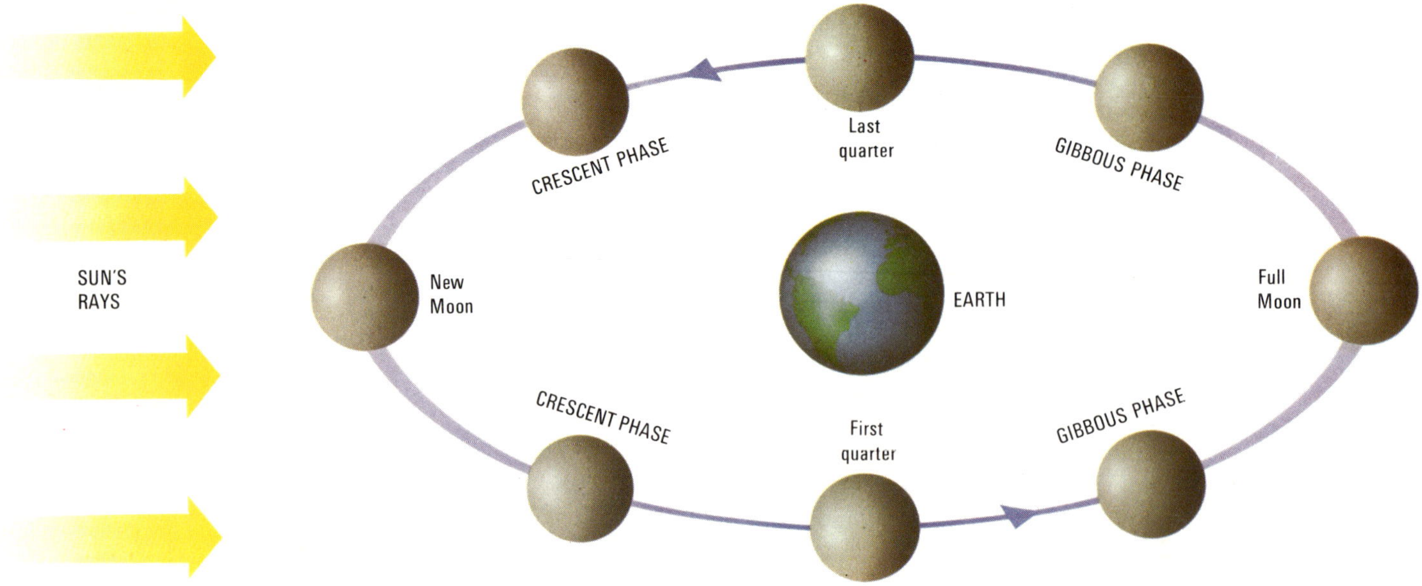

The Moon's Phases

The moon travels around the earth once a month. As it moves, the angle it makes with the sun and earth changes. This gives rise to its *phases*, or changes in appearance. When the moon is closest to the sun, the side facing us is in darkness. We call this a *new moon*. As the moon continues its journey around the earth, the side facing us gradually becomes lit up. The moon passes through crescent, first quarter, and gibbous phases (see diagram). Eventually, it appears as a complete, bright yellow disk. We call this a *full moon*. And the change from new moon to full moon is called *waxing*. Once the moon is full, the reverse process, called *waning*, takes place. Each day, the illuminated area we can see grows less. And, when the moon has become new, once more, the whole sequence repeats.

Although the moon's gravity is weak, it has a noticeable effect on the earth. It pulls on the seas and oceans, causing the

A map of the moon. The fairly flat lowlands on the near side of the moon were once thought to be seas. The rugged highlands are covered with thousands of craters. The far side has no large flat parts.

tides. The height of a tide depends on the positions of the sun, moon, and earth. So the tides vary with the phases of the moon. The highest tides occur at new moon and full moon.

Eclipses

The sun, moon, and earth sometimes lie on the same straight line. When the moon is directly between the sun and earth, an *eclipse of the sun* occurs. And an *eclipse of the moon* occurs when the earth is directly between the sun and moon. For the moon passes right through the earth's shadow.

The moon is eclipsed when the earth's shadow falls on it. A total eclipse occurs if the moon passes through the narrow cone of deep shadow called the umbra. But, if the moon passes only through the penumbra (region of partial shadow), only a partial eclipse occurs. In this case, the earth's shadow never completely covers the moon. The pictures below show the stages of a total eclipse of the moon.

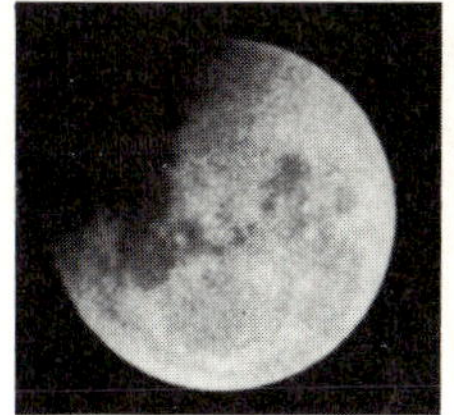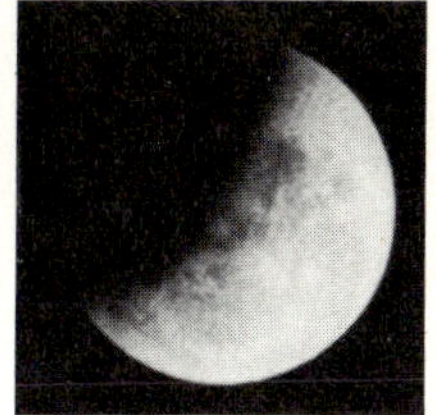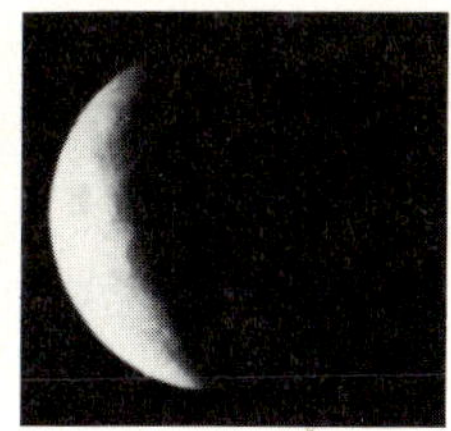

The Lunar Surface

Man's exploration of the lunar surface has revealed much about the moon's history. For the moon has no water or atmosphere to destroy evidence of past events on the surface.

Even a small telescope will pick up many features of the moon's surface. The first person to notice this was Galileo Galilei. In 1609, a year after the telescope had been invented, Galileo used one to study the moon. He soon discovered that the brighter regions were rugged highlands, and that the darker areas were lower and quite flat. At first, astronomers thought that these dark areas were water. So they called them *maria* – Latin for seas. Later observers used more powerful telescopes and found that the moon is, in fact, dry. But, to this day, astronomers still call the moon's lowlands "seas".

The moon's surface has thousands of circular craters. Most of these were probably formed by meteorites – rocks from outer space. Moon craters vary in size. The largest crater, named Clavius, is 233 kilometers (145 miles) across, and more than 3·7 kilometers (2·3 miles) deep.

Craters formed by meteors have a raised rim. This consists of debris blasted from below the ground by the meteor. Some material usually falls in separate lumps, some distance from the main crater. As a result, hundreds of smaller craters may be formed around the main one. Material "splashed" from a main crater may also cause *crater rays*. These are the long, light-colored glassy streaks that radiate from some of the larger craters. The best example of crater rays can be seen around the crater Copernicus.

Meteorites approaching the earth become extremely hot because of friction

Above: Apollo 8 astronauts took this picture of the earth in the moon's sky.

Right: A close view of the far side of the moon. The photograph was taken from an orbiting spacecraft. The large crater is about 80 kilometers (50 miles) across.

Exploring the Moon

What is on the far side of the moon? This question had puzzled astronomers for centuries. For the moon rotates on its axis in just over 27·3 days – exactly the same time as it takes to orbit the earth. So one side of the moon is always hidden from the earth. The far side of the moon remained a mystery until 1959. In that year, the Russian space probe *Lunik 3* took the first photographs of the far side of the moon. The pictures amazed astronomers, for they showed that the back of the moon had no large "seas". Just why the two sides of the moon are so unlike is still not clear.

In preparation for exploring the moon, Russian and American scientists sent many probes to the moon in the early 1960s. And, by 1969, the United States had developed the necessary equipment and techniques for landing men on the moon. On the morning of July 21, 1969, astronauts Armstrong and Aldrin stepped onto the lunar surface. Experiments carried out during this and later moon flights have taught us much about the moon. And the study of moon rocks may tell us how the moon and earth were formed.

with the earth's atmosphere. So they usually burn up and rarely reach the ground. But the moon has no atmosphere to protect it from meteorites. This could account for its many large craters. But not all lunar craters were formed in this way. Some are volcanic craters like those on earth.

Above: American astronaut Harrison Schmitt on the moon in December, 1972. He is collecting samples of soil to take back to earth.

Below: A thin slice of moon rock lit by polarized light and viewed under a microscope.

Mercury and Venus

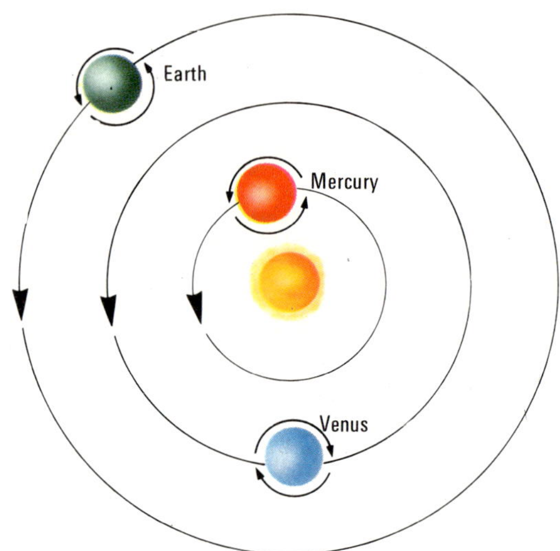

Left: Most planets spin on their axis in the same direction as the earth. But Venus spins the other way.

Mercury and Venus are closer to the sun than the earth is. They both have scorching daytime temperatures. But they differ from each other in many ways.

Mercury and Venus are the planets closest to the sun. Because they are so close to the sun, both planets have scorching daytime temperatures. But they have little else in common. The diameter of Venus is about 2·5 times that of Mercury. So its volume is about 16 times as great. The tiny planet Mercury has almost no atmosphere, and is almost like the moon. But Venus is surrounded by a vast, dense atmosphere that blocks our view of its surface. Venus also turns in the opposite direction to most other planets.

Mercury – the Hot Midget

Mercury, the smallest of the main planets, is not much bigger than our moon. And, like the moon, it shows phases. Mercury is like the moon in other ways too. Its surface is covered with craters. And its temperature varies greatly from day to night. During the day, the surface temperature on Mercury probably reaches about 400°C – hot enough to melt lead. But, at night, the temperature drops to around –200°C – cold enough to turn oxygen gas into liquid. As on the moon, the extremely cold nights are due to the fact that there is no "blanket" of atmosphere to keep in the heat.

Most planets have orbits that are almost circular. But Mercury's orbit is like a long oval. The planet's distance from the sun varies from 47 to 69 million kilometers (29 to 43 million miles).

Left: Venus, photographed from Mariner 10 spacecraft in February 1974, from a distance of 720,000 kilometers (450,000 miles).

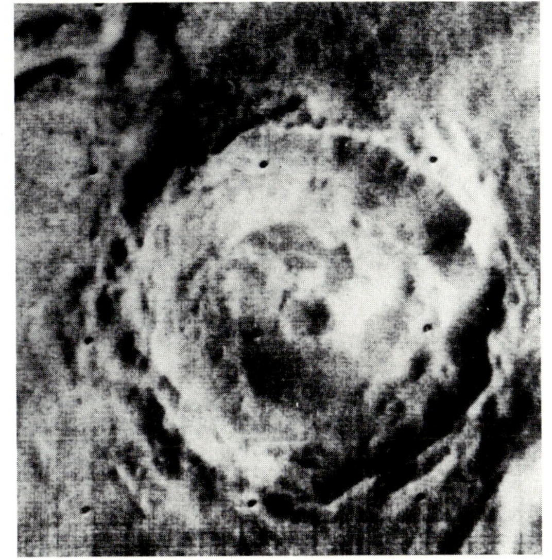

1877, when the Italian astronomer Giovanni Schiaparelli announced that he had discovered "canals" on Mars. Many people thought that these straight markings must be irrigation systems built by Martians to supply water to the desert regions. But later, better observations showed that this and many other ideas about Mars were quite wrong. The "canals", for example, are now thought to be mountain ridges or chains of craters. And the idea that the large, dark areas on Mars were water had been abandoned by the year 1900. For the Martian atmosphere was known to have little water vapor. So most astronomers thought that the dark areas were, in fact, some kind of simple vegetation. This theory fitted in with the fact that, in the Martian spring, these areas grew bigger. Vegetation would be expected to spread in the same way when the weather grew warmer. But recent measurements have supported yet another theory for these changes in the Martian surface. It now seems likely that certain minerals, as they warm up in spring, release water vapor and change color. When the colder weather returns, they absorb water again and go back to their former color.

Another thing that happens during the Martian spring is the shrinking of the planet's polar caps. During winter, the southern cap may spread half way to the equator. But, by mid-summer, it has sometimes vanished altogether. Measurements have shown that the Martian atmosphere has much more water vapor in the polar regions than at the equator. This seems to confirm that frozen water, present in the caps, starts to melt and then evaporate in spring. The polar caps may also contain some dry ice – solid carbon dioxide. This substance, in gas form, certainly makes up the main part of the Martian atmosphere.

Conditions on Mars cannot support the many forms of life that exist on earth. But could simple living things exist in the Martian soil? This was among the many questions that American scientists hoped to answer when they put two unmanned *Viking* landing craft on Mars in the summer of 1976. The two machines scooped soil from the Martian surface and then carried out various tests on it. Results sent back to earth showed that the Martian soil does behave as if it contains living things. But today's scientists are very cautious about drawing conclusions from the results of their experiments. And we may have to wait until man manages to bring back samples of the Martian soil. Then we will know, for certain, whether or not life exists on Mars.

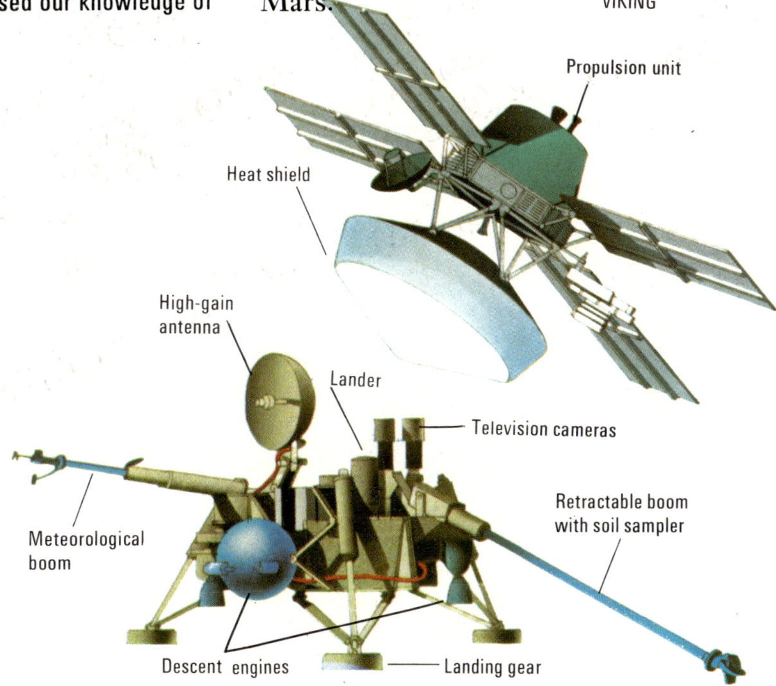

The Outer Planets

Beyond Mars lie the giant planets Jupiter, Saturn, Uranus, and Neptune. And still farther out is the tiny planet Pluto. Another small planet, as yet undiscovered, may lie beyond Pluto.

Above: Jupiter, as seen from the earth, showing bands of clouds. The Great Red Spot, probably a never-ending storm, can be seen clearly. The dark spot above it is the shadow of one of Jupiter's 13 satellites.

Left: To observers on earth, the appearance of Saturn changes throughout its 30-year journey around the sun. The planet's rings are so thin they can hardly be seen when they are edge-on to us.

Jupiter – the Giant Planet

Jupiter is the largest known planet. Its volume is more than 1,300 times that of the earth. Jupiter also has more moons than any other planet – 13 in all. Like the earth, it has a magnetic field. But its strength is ten times as strong as the earth's magnetic field.

On the surface of Jupiter, the temperature is around $-146°C$, with little variation from day to night. Around the giant planet is an atmosphere made up mainly of hydrogen. Small amounts of several other gases are present too. These include ammonia, which forms white clouds.

At its closest, Jupiter is about 640 million kilometers (400 million miles) from earth. But, because it is so large, and its clouds reflect a lot of sunlight, it can be seen clearly with the naked eye. Jupiter is an interesting planet to look at. Its swift rotation pulls its clouds into bands around the equator. The most prominent feature of the planet's atmosphere is its Great Red Spot. This huge oval mark measures about 40,000 kilometers (25,000 miles) by 13,000 kilometers (8,000 miles). It is thought to consist of a swirling mass of gases in a never-ending storm. The spot drifts slowly across the planet, and changes in brightness from time to time. Occasionally, it disappears altogether.

Because of its cloud cover, we do not know what Jupiter's surface looks like.

Saturn – the Ringed Planet

Saturn was the most distant planet known to the ancient astronomers. It is the second largest planet in the solar system, with a diameter of 120,000 kilometers (74,700 miles). Like Jupiter, Saturn has a hydrogen-rich atmosphere with bands of cloud that reflect sunlight well. And Saturn's beautiful ring system adds to its brilliance. The rings are thought to consist of countless dust particles, small rocks, or lumps of ice.

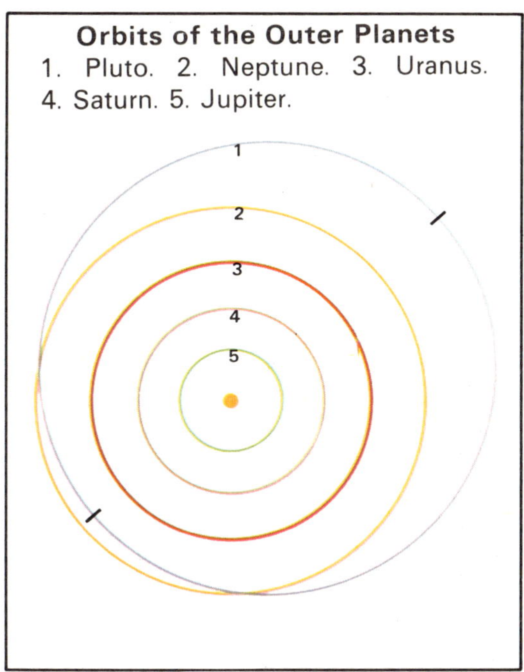

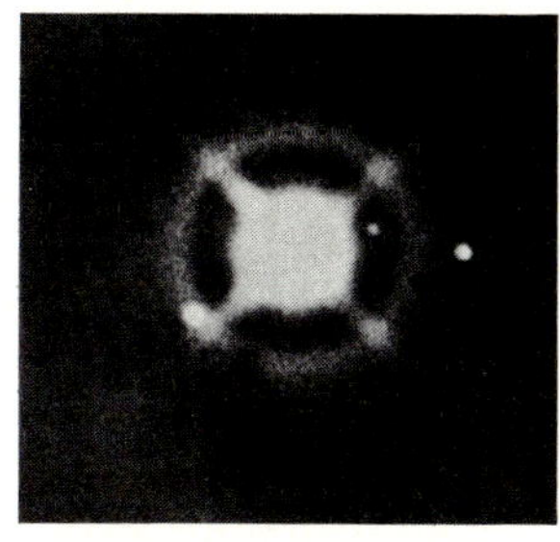

Above: Uranus and its five satellites.

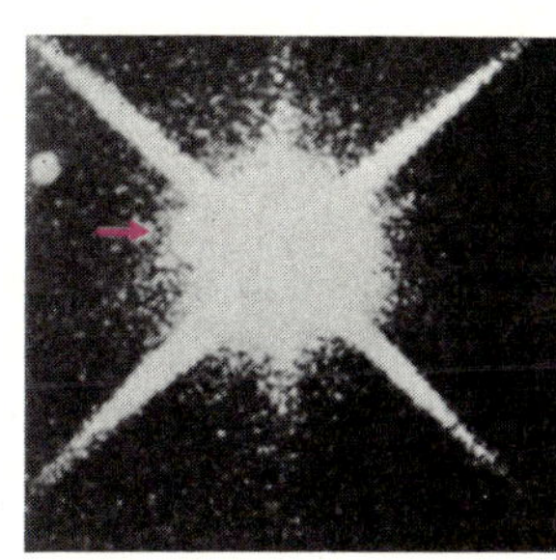

Above: Neptune with its two satellites, Triton and Nereid (arrowed).

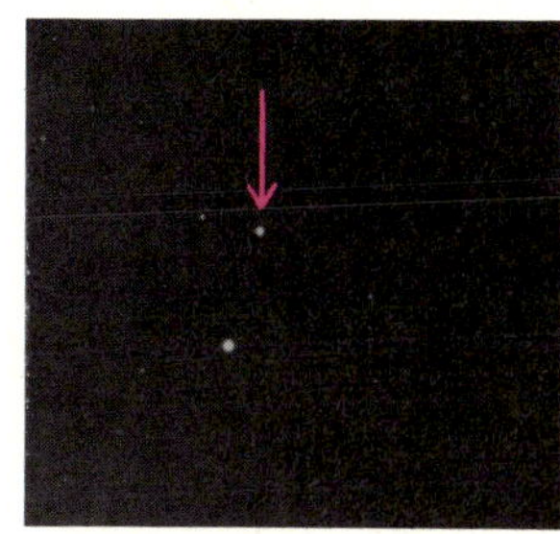

Above: Because Pluto (arrowed) is so far away, and so small, it cannot be seen with the naked eye. When viewed through a powerful telescope, the planet resembles a faint star.

Left: Orbits of the outer planets (not to scale). Pluto's orbit is eccentric (off center), so that Pluto is sometimes closer to us than is Neptune.

Because of its cloud cover, little is known about Saturn. But its structure may resemble that of Jupiter. Saturn has 10 known moons. The last of these was discovered in 1966.

Uranus and Neptune – the Twin Planets

Uranus and Neptune are similar in many ways. Both have a diameter about four times that of the earth. And both are covered by dense cloud – like Jupiter and Saturn. But, in the case of Uranus and Neptune, the most common gas in the atmosphere is methane. Occasionally, Uranus is just visible to the naked eye. But Neptune can never be seen without a telescope. Both planets are orbited by moons. Uranus has five moons, while Neptune has two.

Pluto and Beyond

Pluto, the second smallest of the known planets, has a diameter of only 6,400 kilometers (3,980 miles). The average distance of Pluto from the sun is greater than that of any other planet. But Pluto's orbit is off-center, and crosses Neptune's orbit. As a result, Pluto sometimes travels inside Neptune's orbit – as it is doing today. Pluto will be at its closest to us in 1989. It will then start to move away again and will be at its farthest in the year 2113.

Little is known about Pluto, but it is probably composed of extremely dense material. Some astronomers think that Pluto was once a moon of Neptune.

In 1972, American astronomers predicted that another planet would be found beyond Pluto – at a distance of about 9,660 million million kilometers (6,000 million million miles) from the sun. The gravitational attraction of such a planet would account for slight variations in the orbits of some comets. In a similar way, the existence of Pluto was predicted 15 years before its discovery, after variations in Neptune's orbit had been observed.

33

Space Junk

Besides the main planets and their satellites, many other bodies orbit the sun. These include comets, meteoroids, and asteroids.

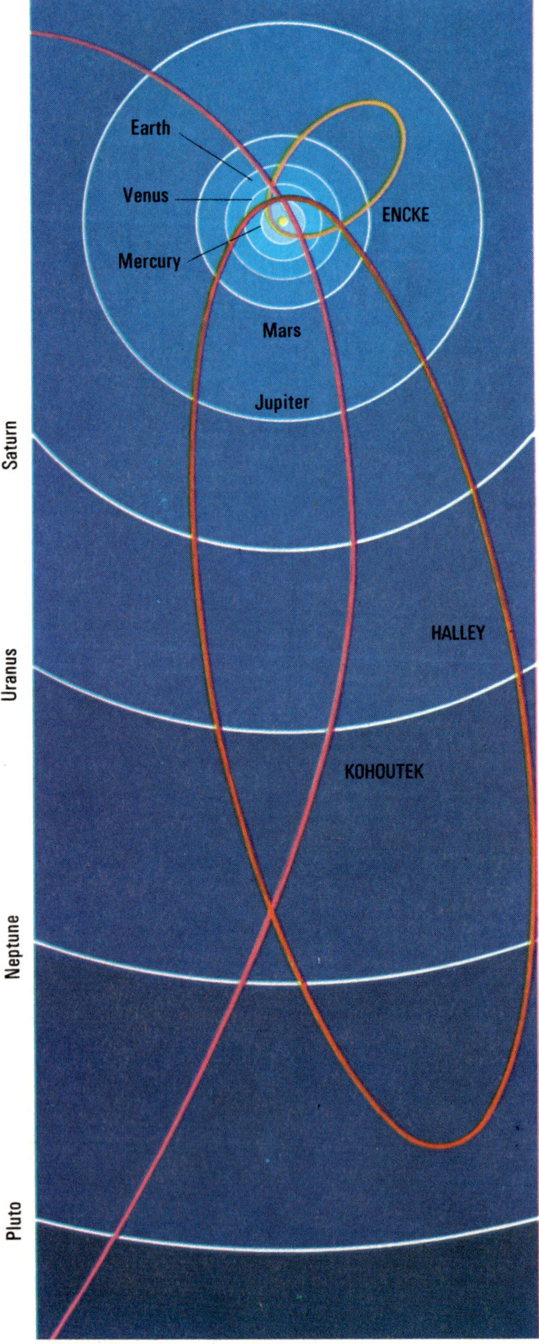

Below: Orbits of the comets Encke, Halley, and Kohoutek. Encke's comet orbits the sun once every 3·3 years. Halley's comet takes about 76 years. But Kohoutek's comet takes 75,000 years to complete its journey.

Above: Part of the Bayeux Tapestry, which shows scenes from the Norman Invasion of England in 1066. Some of the people are pointing to the bright comet visible at that time. Centuries later it was named in honor of the British astronomer Edmund Halley.

Below: Kohoutek's comet passed within 120 million kilometers (75 million miles) of the earth on January 15, 1974.

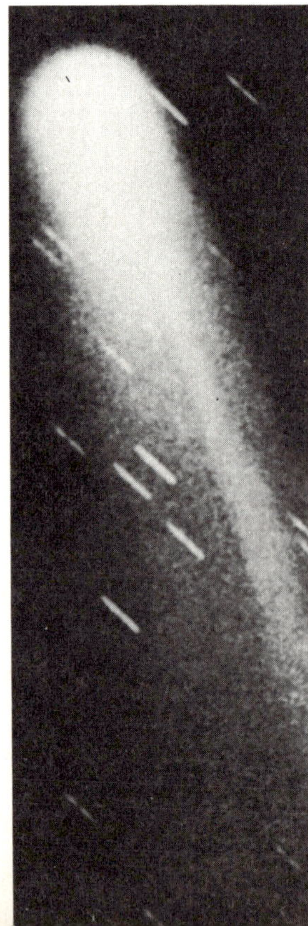

Comets

Comets are regular visitors to our part of the solar system. They are probably made of dust, ice, and frozen gases. Comets glow and develop a "tail" of glowing gases when they pass near the sun. Like the planets, comets orbit the sun. But whereas the orbits of most planets are almost circular, comets' orbits are very long. Although comets regularly pass close to the sun, they usually spend most of their time among, or beyond, the orbits of the outer planets.

Short-term comets orbit completely within our planetary system. Halley's comet, for example, never quite reaches the orbit of Pluto. Every 76 years, Halley's comet passes close to the sun and can be seen with the naked eye. Encke's comet has a shorter period (orbit time) than any other comet. It passes the sun every 3·3 years.

Long-term comets travel beyond Pluto and are rarely seen. Kohoutek's comet, which passed within 120 million kilometers (75 million miles) of the earth in 1974, is a long-term comet. It will not return for another 75,000 years.

Astronomers have estimated that there are about 100,000 million comets altogether. But many of them pass close to the sun only once in several million years.

The brightest comets are visible in broad daylight. And their tails may stretch half-way across the sky.

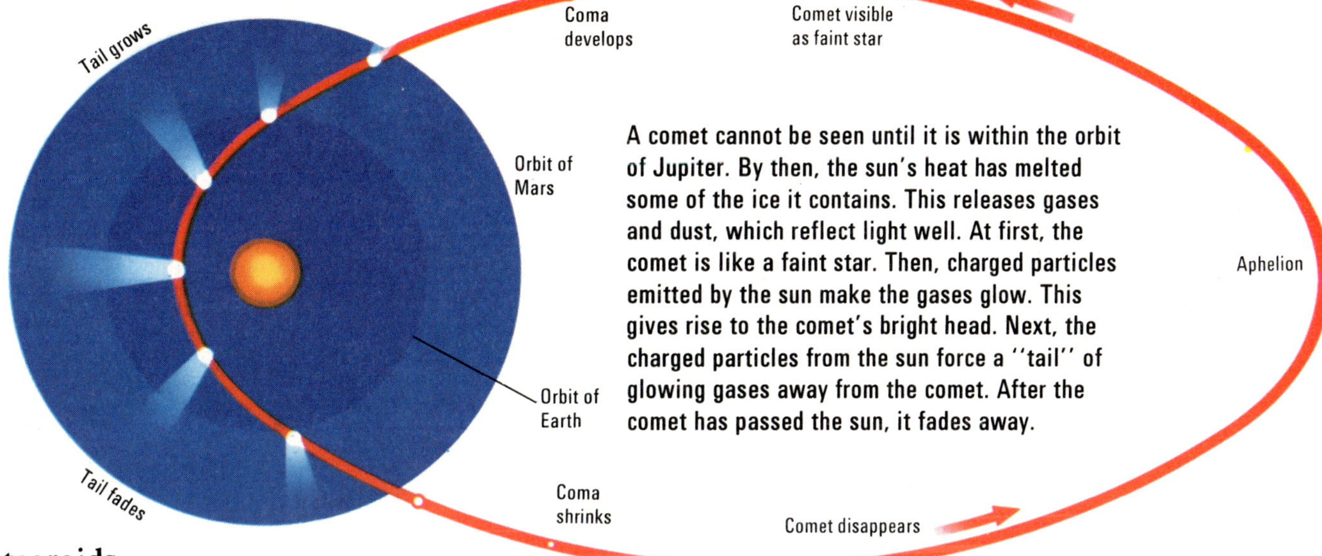

A comet cannot be seen until it is within the orbit of Jupiter. By then, the sun's heat has melted some of the ice it contains. This releases gases and dust, which reflect light well. At first, the comet is like a faint star. Then, charged particles emitted by the sun make the gases glow. This gives rise to the comet's bright head. Next, the charged particles from the sun force a ''tail'' of glowing gases away from the comet. After the comet has passed the sun, it fades away.

Meteoroids

Meteoroids are solid bodies that have broken away from comets. The smallest meteoroids are no bigger than grains of sand, while the largest weigh several tons. Meteoroids are always entering the earth's atmosphere. Friction with the atmosphere usually heats them up so much that they burn up before reaching the ground. This gives rise to bright streaks called *meteors*, *shooting stars*, or *falling stars*, in the night sky. A *meteor shower* is a spectacular display that occurs when the earth passes through a cloud of meteoroids.

Large meteoroids sometimes survive their journey through the atmosphere and smash into the ground. We call these rocks from outer space *meteorites*. The smallest meteorites are like pebbles. But the largest weigh many tons and blast out huge craters when they hit the ground. Fortunately, such meteorites are very rare.

Right: Most of the asteroids, or minor planets, move around the sun in a wide belt between the orbits of Mars and Jupiter. The largest asteroid, named Ceres, is of this type. But, as shown here, a few asteroids move outside the belt.

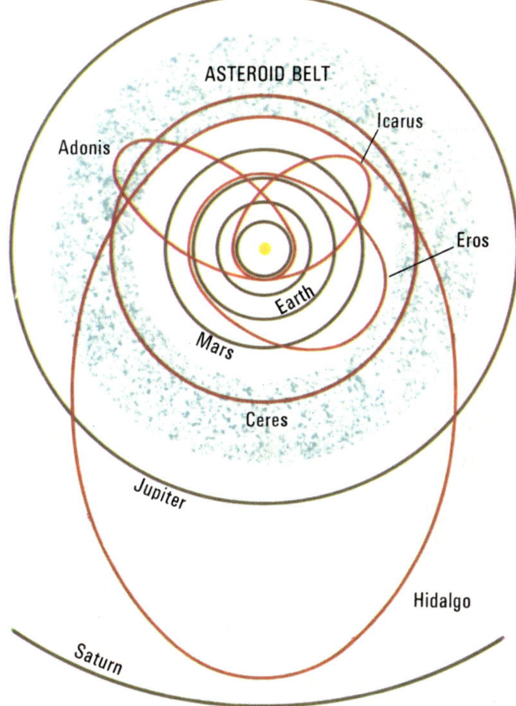

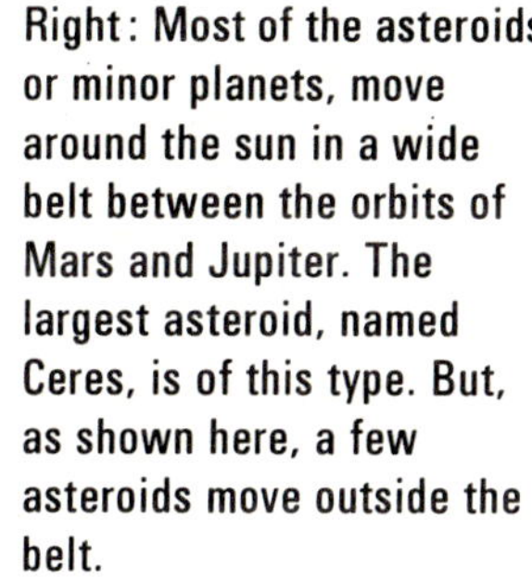
Left: Part of the floor and walls of the Barringer, or Coon Butte meteorite crater in Arizona. This, the largest known meteorite crater on earth, measures about 1·3 kilometers (0·8 miles) across; and is 175 meters (575 feet) deep. It was probably produced between 25,000 and 50,000 years ago by a meteorite no more than 80 meters (260 feet) across.

Asteroids – the Minor Planets

Like meteoroids, asteroids are rocks in space. But asteroids were probably formed at the same time as the main planets. Most asteroids orbit the sun in a near circular path. But some asteroids have eccentric orbits (see diagram). Few asteroids are greater than 160 kilometers (100 miles) in diameter.

Telescopes

The invention of the telescope in the 1600s led to great advances in astronomy.

Above: An astronomer using the 508-cm (200-in) Hale reflecting telescope at Mount Palomar Observatory.

Above: A replica of Newton's reflecting telescope. Below left: The 100-cm (40-in) refracting telescope at Yerkes Observatory, near Chicago.

Optical Telescopes

There was a great change in astronomy in the early 1600s. Before that time, astronomers could do little but study the positions and movements of the stars. Then, in 1608, the Dutch spectacle maker Hans Lippershey invented the telescope. At last, astronomers could observe the moon, sun, and close planets in detail. And thousands of stars, invisible to the naked eye, could be seen for the first time.

In 1609, Galileo Galilei became the first astronomer to use a telescope. Although it was a weak, crude instrument, Galileo made many important discoveries with it. Galileo's telescope was known as a *refracting* telescope. It gave a magnified image of distant objects by refracting (bending) light.

A refracting telescope has two lenses – an *objective* and an *eyepiece*. The objective lens forms an image of the object being studied. And the eyepiece magnifies this image.

One problem of early telescopes was that the image produced suffered from color distortion. Fringes of red and blue were formed around the edge of the image. In the 1660s, Isaac Newton discovered why this was. White light, he found, could be split up into

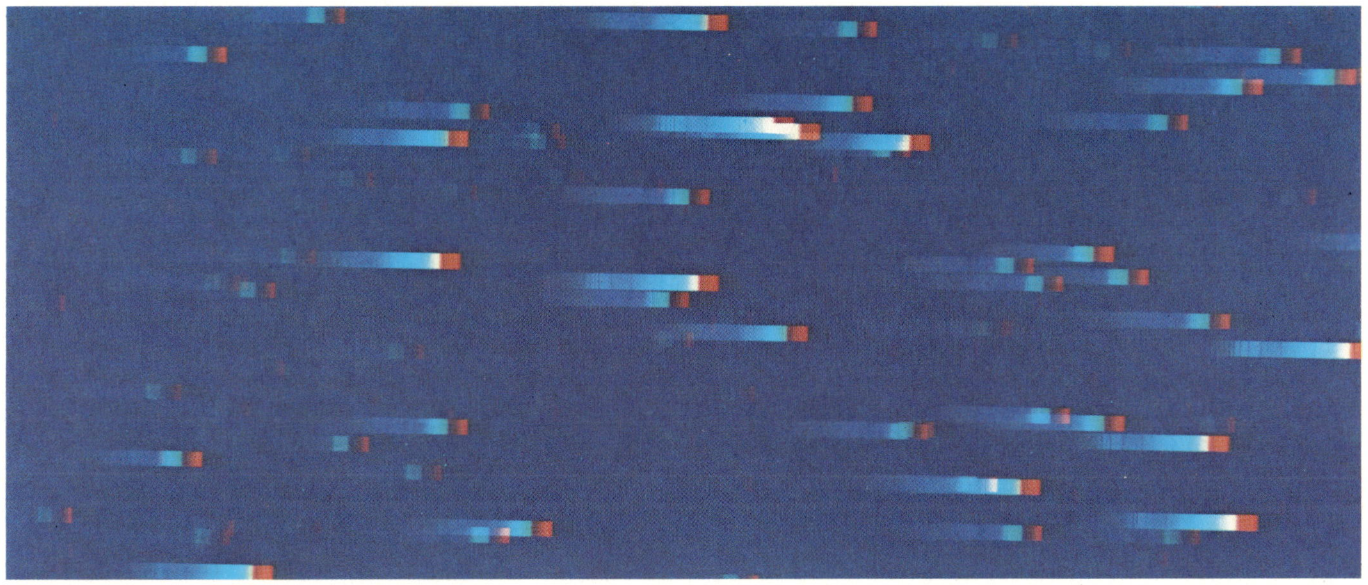

Below: A glass prism splits up a beam of light into a spectrum of colors.

Above: The Hyades star cluster, photographed through a telescope and a prism. The prism has split up the light from each star into a spectrum of colors.

a *spectrum*, or band of colors, by passing it through a glass prism. Telescope lenses were doing the same thing. This discovery led Newton to build the first *reflecting* telescope. Color distortion was greatly reduced because the telescope had a curved mirror instead of an objective lens. The curved mirror reflected light from the object onto a plane mirror. It then passed into an eyepiece lens in the side of the telescope tube. The reflecting telescope soon became popular with astronomers.

In the 1750s, an instrument maker called John Dollond discovered how to make lenses with low color distortion. Each of Dollond's lenses was made up of two separate lenses stuck together. These two lenses were made of different types of glass. Although one lens split up incoming light into various colored rays, the other lens joined them together again. As a result, the images obtained were almost free from color distortion. Dollond soon started making his *achromatic* (without color) lenses. Their use greatly improved the refracting telescope. So this kind of telescope became popular with astronomers once more.

Several different types of reflecting and refracting telescopes are used by astronomers today. The largest telescopes are of the reflecting type. For large lenses tend to bend under their own weight and give a distorted image. A lens can be supported only around its rim, but a mirror can be supported over the whole area of its back surface. Bending is thus prevented, and there is no distortion.

Photographs and Optical Measurements

In an observatory, electric motors slowly move the telescope so that it follows, or tracks, the heavenly body under investigation. Accurate tracking of the object is very important when photographs are being taken through the telescope. The light from stars

37

Continuous Spectrum

The glowing gases inside stars are under very great pressure and give off white light with a continuous spectrum (above). A glowing gas at low pressure yields a bright-line spectrum (below), which is characteristic of that gas.

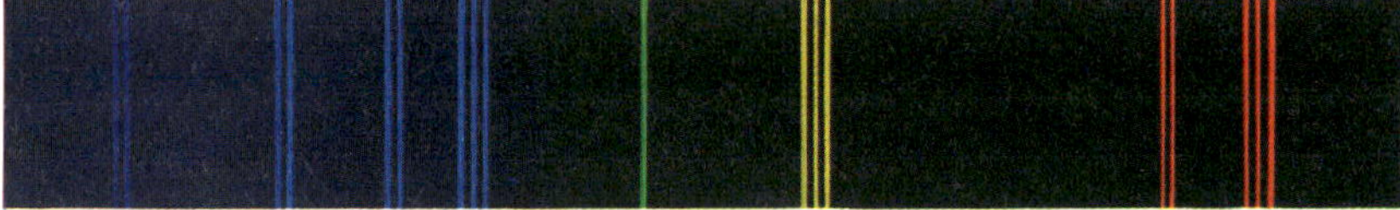

Emission Spectrum

When the white light from a star's interior passes through the cooler, lower-pressure gases of the star's atmosphere, dark lines appear in the spectrum (below). They appear in the same positions as would the lines in the emission spectra of those gases.

Absorption Spectrum

Top left: The radio telescope at Effelsberg, near Bonn, West Germany. The telescope's concave dish measures 100 m (328 ft) across.

Above: The control room from which astronomers operate the Effelsberg radio telescope.

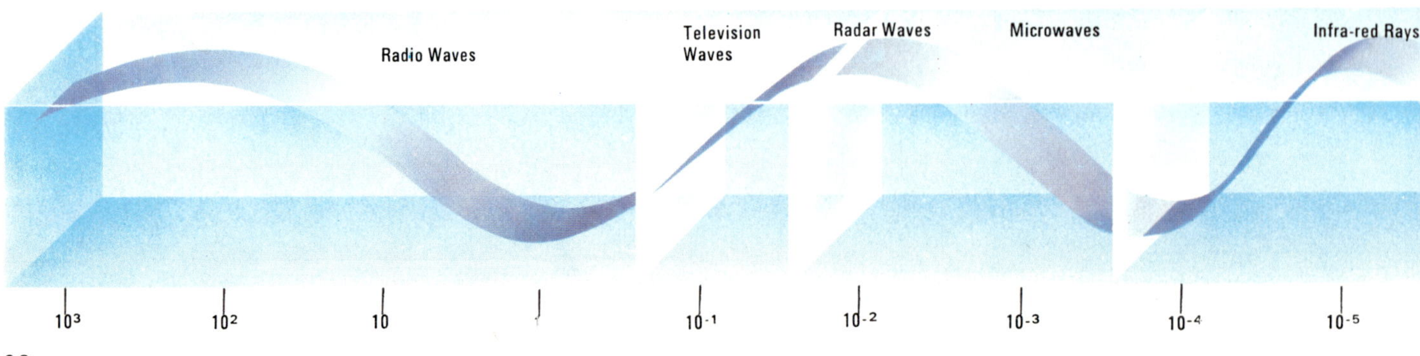

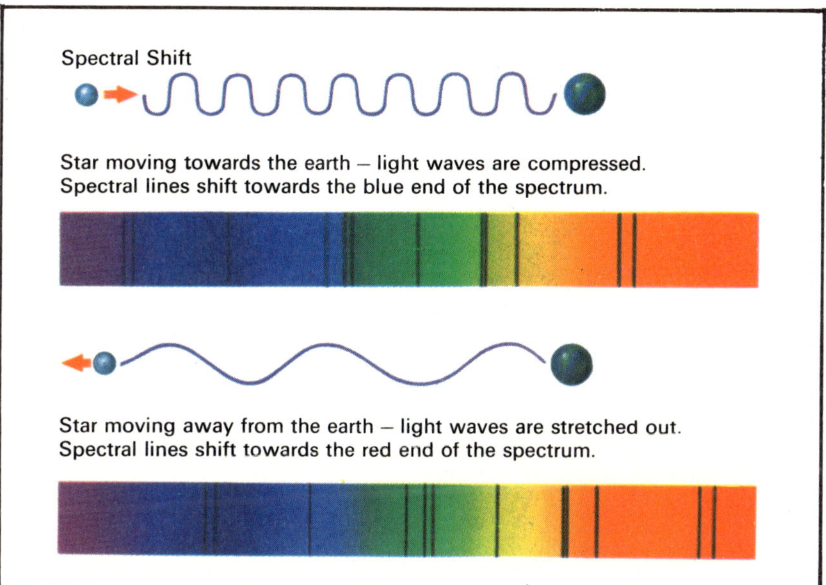

Below: The electromagnetic spectrum, or range of electromagnetic waves. Electromagnetic waves of certain frequencies can be seen by the eye and are called light waves. Radio waves have lower frequencies and do not stimulate the eye. But these waves can be detected by radio equipment.

A high frequency corresponds to a short wavelength. And a low frequency corresponds to a long wavelength.

is very dim. So the photographic film has to be exposed to it for a long time to record an image. If the telescope did not track accurately, the photograph would be blurred.

Even the best telescopes show the stars as tiny points of light. But astronomers can learn much about a star with an instrument called a *spectroscope*. It splits the starlight into a spectrum of colors. From this band of colors astronomers can work out what the star is made of, how it moves and other things about it.

Invisible Astronomy

Stars give out energy in the form of waves that can travel through space. Scientists call them *electromagnetic waves*. Light waves are just one form of this radiation. Other forms include radio waves, infrared (heat) rays, ultraviolet rays, and X-rays. Astronomers use special instruments to study these invisible rays.

The earth's atmosphere absorbs most electromagnetic waves, so they are prevented from reaching the ground. But they can be studied from balloons, rockets, and satellites high above the earth. Only light waves and radio waves reach the ground without much reduction in strength. So astronomers talk about two imaginary "windows" in the atmosphere. Through these windows they can study the universe from observatories on earth. Waves passing through the radio window are picked up by the huge antennas of *radio telescopes*. These are usually dish-shaped.

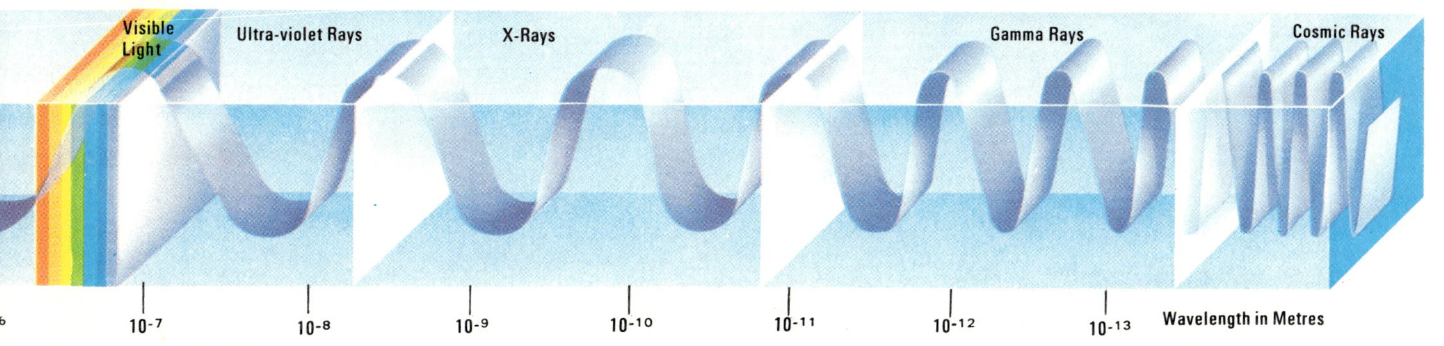

Exploring Space

Techniques for exploring space have progressed at an amazing speed. Less than eight years after Yuri Gagarin became the first spaceman, Neil Armstrong set foot on the moon.

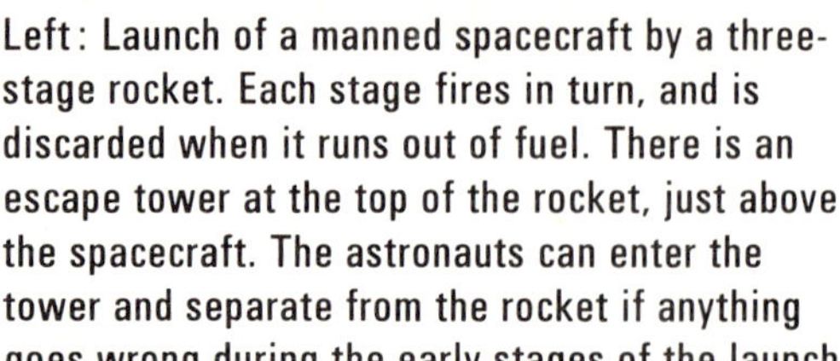

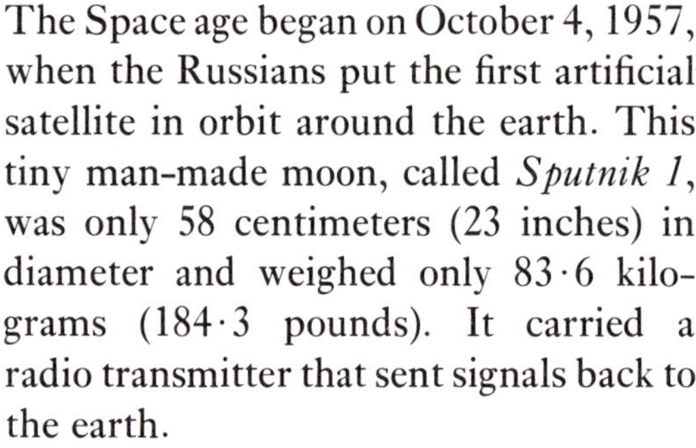

Left: Launch of a manned spacecraft by a three-stage rocket. Each stage fires in turn, and is discarded when it runs out of fuel. There is an escape tower at the top of the rocket, just above the spacecraft. The astronauts can enter the tower and separate from the rocket if anything goes wrong during the early stages of the launch.

The Space age began on October 4, 1957, when the Russians put the first artificial satellite in orbit around the earth. This tiny man-made moon, called *Sputnik 1*, was only 58 centimeters (23 inches) in diameter and weighed only 83·6 kilograms (184·3 pounds). It carried a radio transmitter that sent signals back to the earth.

Since 1957, space technology has advanced rapidly, and hundreds of artificial satellites now circle the earth. Some are used for communications, weather forecasting, or studying the earth. Others photograph the heavens. From their position well above the atmosphere, these space observatories get much clearer pictures of the stars and planets than could be obtained from the earth. Artificial satellites have also been put into orbit around the moon and around Mars. And unmanned space probes have soft-landed on Mars. But the greatest achievement of all has been manned spaceflight.

Man in Space

On April 12, 1961, the Russian Yuri Gagarin became the first man to circle the earth. The first American to do this was John Glenn, in February 1962. But these great achievements were only the first steps. Soon, highly advanced spacecraft and launching rockets had been developed. And, in July 1969, the American astronaut Neil Armstrong became the first man to set foot on the moon.

Meanwhile, the Russians were building the first space laboratory, named *Salyut*. They put this in orbit around the earth in April 1971. And, a few days later, a manned *Soyuz* satellite joined the orbiting laboratory. The *Soyuz* cosmonauts latched the two space vehicles together. Then they climbed through into *Salyut*, where they spent a few hours before returning to earth in *Soyuz*. And, in May 1973, the Americans launched their own space laboratory, called *Skylab*. In just under a year, three teams of astronauts visited *Skylab*. Their work included studying the earth, sun, and the comet Kohoutek.

Spacecraft are launched by multi-stage rockets. When one stage runs out of fuel, it separates from the rest of the vehicle. Then the next stage of the rocket fires. Disposing of used rockets makes space travel extremely expensive. In the 1980s, the U.S.A. intends to start operating a re-usable launching vehicle called a space shuttle. Only a cheap fuel tank will be lost on each mission.

Above: In the summer of 1975, Russian and American spacecraft, in earth orbit, linked up for the first time.

Right: Skylab, the American space laboratory, in orbit around the earth in 1973.

Above: On board Skylab, Dr Joseph Kerwin inspects Charles Conrad's mouth. In the weightless conditions of space, men float around and have no sense of ''up'' or ''down''.

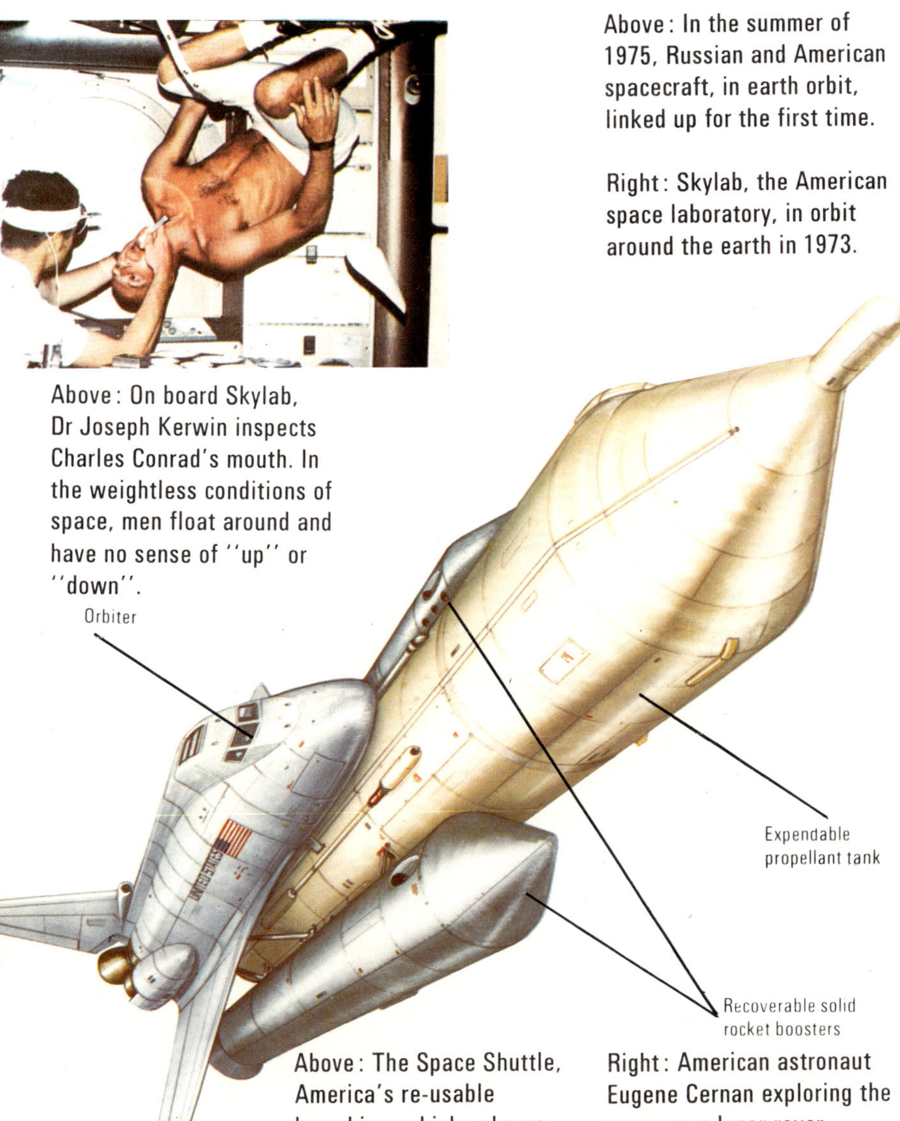

Above: The Space Shuttle, America's re-usable launching vehicle, shown ready for take-off. In the 1980s, it will launch up to 70 spacecraft each year.

Right: American astronaut Eugene Cernan exploring the moon on a lunar rover vehicle in December, 1972.

Glossary of Terms

Aberration (1) A shift in the apparent position of a heavenly body, caused by the movement of the observer with the earth. (2) Chromatic aberration. The formation, by a lens, of unwanted color fringes around the image produced. (3) Spherical aberration. Blurring of the image formed by a lens or curved mirror. Different parts of the lens or mirror form images at slightly different positions.

Apogee The apogee of a satellite is the point at which it is farthest from the body it is orbiting.

Aurora A display of colored light in the sky near the earth's north or south pole. Charged particles from the sun are attracted to the poles. When the particles enter the upper atmosphere, they make the gas there glow. In the northern hemisphere, this effect is called the aurora borealis, or northern lights. In the southern hemisphere, it is known as the aurora australis, or southern lights.

Big-bang theory The theory that the universe started with an enormous explosion. This theory is now accepted by most scientists.

Coma The light given off from around the solid part of a comet when it is near the sun.

Constellation A pattern of stars. The stars in a constellation are not usually grouped together. They merely lie in the same general direction as viewed from the earth.

Corona The sun's glowing outer atmosphere, visible only during a total eclipse of the sun. The corona then appears as a pearly white halo.

Cosmic rays Charged particles reaching the earth from outer space.

Eclipse (1) Solar. A cutting off of sunlight, caused by the moon passing directly between the sun and earth. (2) Lunar. A cutting off of moonlight, caused by the earth's shadow falling on the moon.

Electromagnetic radiation Waves of energy that travel through space at 300,000 kilometers per second (186,000 miles per second). Electromagnetic radiation includes radio waves, infrared (heat) rays, visible light, ultraviolet rays, X-rays, gamma rays, and cosmic rays.

Escape velocity The speed a rocket must reach in order to escape from the gravitational attraction of a planet or other body. From earth, the escape velocity is about 40,000 kilometers per hour (25,000 m.p.h.). From the weaker gravitational attraction of the moon, the escape velocity is about 8,500 kilometers per hour (5,300 m.p.h.).

Falling star See: *Meteor.*

Inferior planets Planets within the earth's orbit.

Light-year A unit of distance used in astronomy. One light-year, the distance that light travels in one year, is equal to 9·5 million million kilometers (5·9 million million miles).

Magnitude A measure of the brightness of a star or other heavenly body.

Meteor A sudden, bright streak in the sky, caused by a *meteoroid* entering the earth's atmosphere and burning up. Meteors are also called shooting stars, or falling stars.

Meteorite *Meteoroids* that survive their passage through the earth's atmosphere and reach the ground.

Meteoroid A small body that may enter the earth's atmosphere and give rise to a *meteor* trail. The largest meteoroids weigh several tons, but the smallest are no bigger than grains of sand. Probably more than one million meteoroids enter the earth's atmosphere each day. Occasionally, a larger meteoroid falls to earth as a *meteorite.*

Nebula A cloud of dust and gas, from which stars are formed. Nebulae may be bright or dark. Bright nebulae are illuminated by stars within them, or nearby. Dark nebulae cannot be seen because there are no stars close enough to illuminate them. But dark nebulae can be detected because they prevent our seeing stars behind them. This effect gives rise to apparent gaps in the dense band of stars known as the Milky Way.

Nova A star that suddenly increases in brightness, usually by several thousand times. This flare-up is caused by an explosion in the star. It may make the star visible for the first time, hence the term "nova", meaning "new". After the explosion, the star slowly returns to its former state. See also: *Supernova.*

Orbit The path in which one heavenly body repeatedly moves around another. Gravitational attraction between the two bodies prevents their separation.

Perigee The perigee of a satellite is the point at which it is closest to the body it is orbiting.

Perturbation A change in the motion of a planet, satellite, or comet, caused by the gravitational attraction of another heavenly body. Perturbations in the orbit of Neptune led to the discovery of Pluto.

Phases The various shapes of the bright part of a heavenly body. The moon, Mercury, and Venus show phases when viewed from earth. The phases of a heavenly body are caused by changes in the relative positions of the body, the observer, and the sun.

Precession Circular movement of the axis about which a body rotates. The precession of a heavenly body is similar, in some ways, to the wobble of a spinning top. The earth's axis is always at $23\frac{1}{2}°$ to the plane of the planet's orbit around the sun. But the direction in which the axis points changes continuously. At the moment, the axis points almost exactly to Polaris, which is, therefore, the current *pole star.* But, by about the year 14,000, as a result of precession, the axis will have shifted to point at the star Vega, in the constellation Lyra. Polaris will become the pole star again around the year 27,800.

Pulsar Abbreviation for pulsating star. Pulsars emit short bursts of radio waves at regular intervals. Some pulsars emit pulses of light and X-rays too.

Quasar Quasi-stellar (starlike) sources of *electromagnetic radiation.* The powerful radio waves emitted by some quasars led to their discovery by radio astronomers in the early 1960s. All known quasars emit ultraviolet light. Some emit radio waves and visible light too.

Radiation belts See: Van Allen belts.

Radio astronomy The study of heavenly bodies by reception and observation of the radio waves they emit or reflect.

Radar astronomy The investigation of heavenly bodies by sending radio waves to them and studying the reflected signals. Radar astronomy is particularly useful for studying *meteors* during the daytime. Only extremely bright meteors are visible in the bright, daytime sky. But they reflect radio waves equally well during the day and at night.

Satellite Any body that moves in an *orbit* around another, more massive body. The planets, for example, are satellites of the sun. And the moon is a satellite of the earth. Many other artificial, or man-made satellites orbit the earth too. They are used for such purposes as radio and television communications, and for gathering information on the earth and the weather.

Shooting star See: *Meteor.*

Solar wind A stream of electrically charged particles given off by the sun. Some of these particles enter the earth's atmosphere in the polar regions, where they give rise to *auroras.* See also: *Van Allen belts.*

Sputnik Russian word for "traveling companion", and the name given to the first artificial satellites launched by the USSR.

Sunspot An area on the sun's surface that is about 2,000 °C cooler than its surroundings. Although actually bright, sunspots appear in photographs to be dark in comparison with the rest of the sun. A typical sunspot develops within a few hours and lasts for several months. The number of sunspots present at any time varies over an 11-year cycle. Scientists do not know why sunspots occur. But they are associated with the *solar wind.* At times of maximum sunspot activity, the strong solar wind causes spectacular *auroras* at the earth's poles. And the charged particles making up the wind disrupt radio communications on earth.

Superior planets Planets outside the earth's orbit.

Supernova A large star that ends its life by exploding violently. A typical supernova becomes about 100 million times as bright as the sun during the explosion. Many supernovas have been observed in distant galaxies. But only two are known to have occurred within our own Galaxy.

Transit The passage of one heavenly body across the disk of another, larger one. For example, the planet Mercury can occasionally be seen in transit across the sun.

Van Allen belts Two bands of electrically charged particles surrounding the earth. The particles are kept in place by the earth's magnetic field. The inner belt is at an average height of 3,200 kilometers (2,000 miles) above the earth. And the outer belt is at a height of 16,000 kilometers (9,900 miles).

Zodiac The background of 12 *constellations*, past which the sun, moon, and planets appear to travel. The symbols representing these constellations are called the signs of the zodiac. Astrologers (fortune-tellers) imagine that the constellations of the zodiac somehow affect our lives.

The Constellations

CONSTELLATIONS OF THE NORTHERN HEMISPHERE

*Taurus,
the Bull*

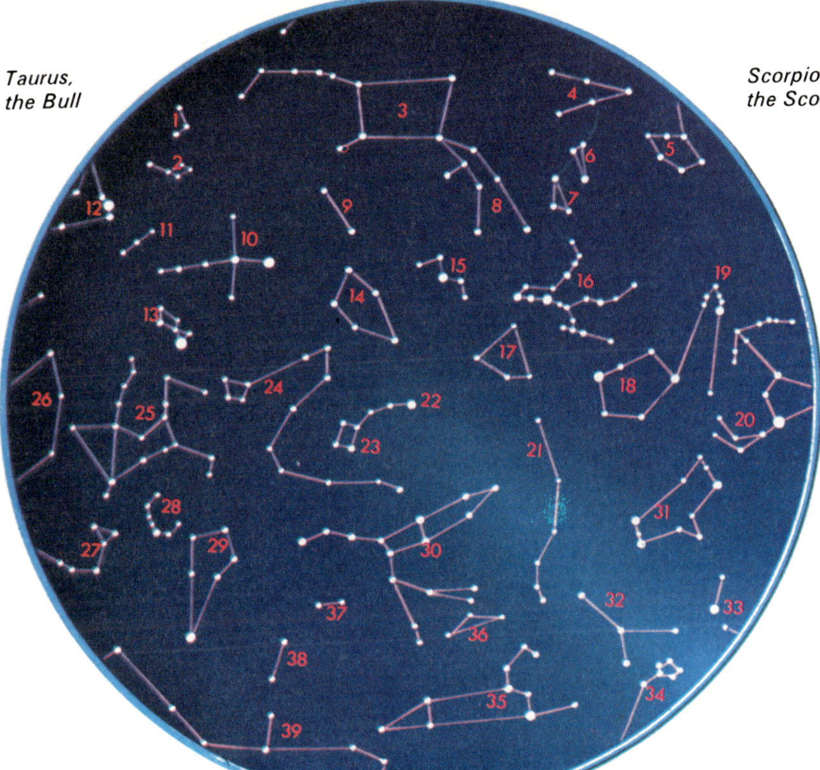

*Scorpio,
the Scorpion*

1 Equuleus, Colt
2 Delphinus, Dolphin
3 Pegasus, Flying Horse
4 Pisces, Fishes
5 Cetus, Sea Monster
6 Aries, Ram
7 Triangulum, Triangle
8 Andromeda,
   Chained Maiden
9 Lacerta, Lizard
10 Cygnus, Swan
11 Sagitta, Arrow
12 Aquila, Eagle
13 Lyra, Lyre
14 Cepheus, King
15 Cassiopeia
   Lady in Chair
16 Perseus, Champion
17 Camelopardus,
   Giraffe
18 Auriga, Charioteer
19 Taurus, Bull
20 Orion, Hunter
21 Lynx, Lynx

22 Polaris, Pole Star
23 Ursa Minor, Little Bear
24 Draco, Dragon
25 Hercules,
   Kneeling Giant
26 Ophiuchus,
   Serpent-Bearer
27 Serpens, Serpent
28 Corona Borealis,
   Northern Crown
29 Boötes, Herdsman
30 Ursa Major,
   Great Bear
31 Gemini, Twins
32 Cancer, Crab
33 Canis Minor, Little Dog
34 Hydra, Sea Serpent
35 Leo, Lion
36 Leo Minor, Little Lion
37 Canes Venatici,
   Hunting Dogs
38 Coma Berenices,
   Berenice's Hair
39 Virgo, Virgin

CONSTELLATIONS OF THE SOUTHERN HEMISPHERE

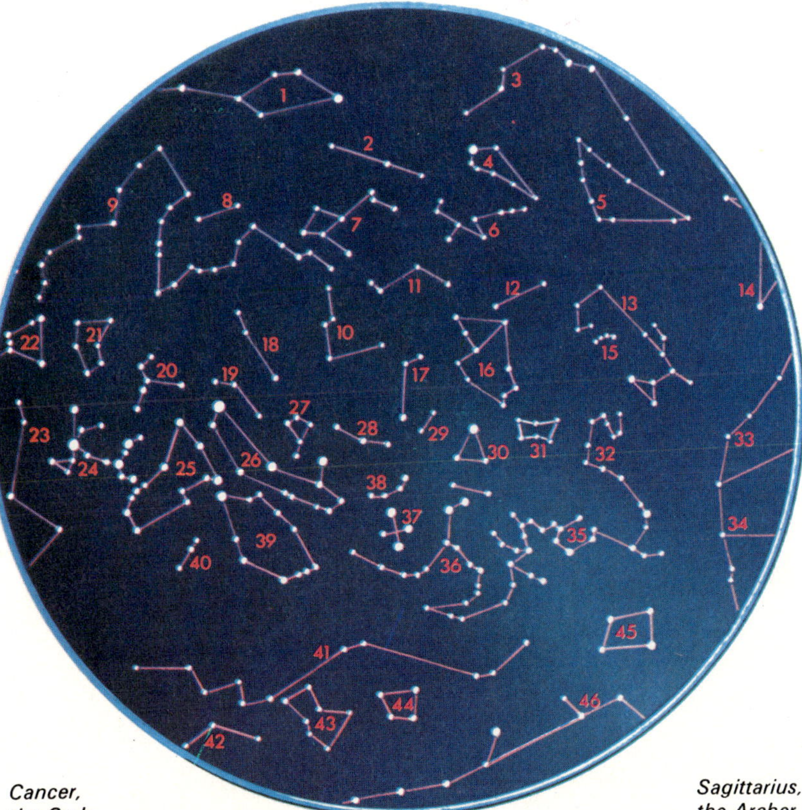

1 Cetus, Sea Monster
2 Sculptor, Sculptor
3 Aquarius,
   Water-Bearer
4 Piscis Austrinus,
   Southern Fish
5 Capricornus, Sea Goat
6 Grus, Crane
7 Phoenix, Phoenix
8 Fornax, Furnace
9 Eridanus,
   River Eridanus
10 Hydrus, Little Snake
11 Tucana, Toucan
12 Indus, Indian
13 Sagittarius, Archer
14 Aquila, Eagle
15 Corona Australis,
   Southern Crown
16 Pavo, Peacock
17 Octans, Octant
18 Dorado, Swordfish
19 Pictor, Painter's Easel
20 Columba, Dove
21 Lepus, Hare
22 Orion, Hunter

23 Monoceros, Unicorn
24 Canis Major, Great Dog
25 Puppis, Poop
26 Carina, Keel
27 Volans, Flying Fish
28 Chamaeleon,
   Chamaeleon
29 Apus, Bird of Paradise
30 Triangulum Australe,
   Southern Triangle
31 Ara, Altar
32 Scorpio, Scorpion
33 Serpens, Serpent
34 Ophiuchus
   Serpent-Bearer
35 Lupus, Wolf
36 Centaurus, Centaur
37 Crux, Southern Cross
38 Musca, Fly
39 Vela, Sails
40 Pyxis, Compass Box
41 Hydra, Sea Serpent
42 Sextans, Sextant
43 Crater, Cup
44 Corvus, Crow
45 Libra, Scales
46 Virgo, Virgin

*Cancer,
the Crab*

*Sagittarius,
the Archer*

Index

Acknowledgements

Page 6 United States Naval Observatory (top), NASA (bottom); 8 British Museum; 9 Istanbul University (top), Roman Picture Library (center); 10 Radio Times Hulton (top, left), Roman Picture Library (top, right), Mansell Collection (bottom); 11 Mansell Collection (top), Radio Times Hulton (center and bottom); 13 United States Naval Observatory; 14 California Institute of Technology; 15 California Institute of Technology (top), United States Naval Observatory (bottom); 17 Lick Observatory; 20 ZEFA/Photri (top), Lockheed Solar Laboratory (bottom); 21 Lockheed Solar Laboratory; 22 NASA (top), ZEFA (center), Solarfilma (bottom); 24 NASA (top), California Institute of Technology (center); 25–28 NASA; 29 NASA (top, right), California Institute of Technology (center), NASA (bottom); 30 NASA (top), California Institute of Technology (bottom); 31 NASA; 32–33 California Institute of Technology; 34 Michael Holford (top), NASA (bottom); 35 American Meteorite Laboratory; 36 California Institute of Technology (top, left), Science Museum (top, left and bottom); 38–39 Max Planck Institute for Radio Astronomy (top, left); 41 NASA.